Concepts in Middle School Mathematics

Dedicated to my team at Astrarka

Foreword

This book covers concepts in Arithmetic, Elementary Algebra, Euclidean Geometry and Plane Trigonometry. The book is organized into four major sections, each dealing with the concepts in a specific topic. The student may use this as a ready reckoner aid in test preparation or with home assignments. The concepts are uncovered gradually – in some sense like LEGO blocks – one piece at the time. We have tried to begin at the very beginning and go to the very end [Alice's Adventures in Wonderland, Lewis Carroll]. The gradation is gentle, yet challenging. Children must understand and explore the music in the language of numbers and symbols – the associated manipulations. This ensures that the child is motivated and stays motivated. We sincerely hope that the student is able to get a good grasp of the subject and the techniques after using this book for meeting his regular academic needs.

Chandramouli Mahadevan.
Astrarka, Bangalore

Contact Information:

Astrarka

225, Block A, AECS Layout, Kundanahalli, Bangalore - 560037, India

Email: support@astrarka.com

Twitter: @astrarka

Web: http://www.astrarka.com

Table of Contents

Concepts in Arithmetic

Concept 0: Good Habits

There are five fundamental principles, or say **good habits** that we would like to emphasize before we commence our discussion on Mathematics.

1. Neatness is conducive to accuracy. Refrain from the temptation to write down something quickly and then scratch the same to make the necessary corrections.

2. One of the weaknesses we find in students while solving word problems is the usage of = sign. This sign has a specific meaning in the world of mathematics. It cannot be used as a way to begin every new line or step in the problem solving process. Use appropriate mathematical signs and symbols. Never use them to mean something vague. = sign is not a space-filler.

3. Spend a second or two to explain how you arrived at a certain step. Several books and references use a statement, such as "it follows from the above statement". We have oftentimes wondered how the expression or equation below follows from the one above. A good explanation is an excellent demonstration of your understanding of the underlying principles.

4. When you are faced with several conclusions during problem solving process, it is a good idea to number the statements or equations. In subsequent steps, you can refer to these conclusions by using the label or the assigned equation number.

5. The easiest of problems attracts the silliest of mistakes. If the problem is easy, motivate yourself to get it right. Do not let over-confidence or carelessness take control of the situation.

Concept 1: Notation

1. Notation: Writing numbers in figures
2. Numeration: Writing numbers in words
3. Each number has two features
   a. Face Value: Does not change. Not positional
   b. Place Value: Does change with position
4. Digits: 0,1,2,3,4,5,6,7,8,9

5. Numeral: Set of digits
6. Place Value System: Start from the extreme right. Move towards the left one step at a time. Each place value is Units, Tens, Hundreds, Thousands and so on
7. In the International Number System, we group 3 digits at a time from the right
8. The place values within the group are Units, Tens and Hundreds
9. The groups from the right are hundreds, thousands, millions. billions, trillions and so on

Concept 2: Comparison of two numbers

1. Fewer the digits, lesser the number
2. If the number of digits in the two numbers are equal, starting from the leftmost digits of the two numbers:
3. Compare the digits
   a. If they are equal; go to the digit to its right
   b. else, larger the digit, larger the number
4. This procedure is iterative

Concept 3: Estimation

Description: We round off to the nearest ten by replacing the digit in the units place using the following rule:

1. If the digit is greater than 5, increase the digit in the tens place. Replace the digit in the units place by 0.
2. If the digit in the units place is less than 5, then we simply replace the digit by 0
3. We can extend the idea to round off a number to its nearest 100s, 1000s or even larger numbers. We simply check if the number to its right is lesser than 5 or greater than 5. Then apply the rule as shown above

Why is rounding off operation required?

When we have to estimate, or make an informed guess or an approximation. We round-off the numbers to the nearest reference number. Population of a town is a good example. We are seldom interested in the actual number. Instead rounding off the number to the nearest 100,000 is typically good enough.

1. Estimating Sum
   a. Round off the individual numbers to the nearest 10s, 100s or 1000s
   b. Add the rounded off numbers

2. Estimating Product
   a. Round off the individual numbers to the nearest 10s, 100s or 1000s
   b. Multiply the rounded off numbers

3. Estimating Quotient
   a. Round off the individual numbers to the nearest 10s, 100s or 1000s
   b. Divide the rounded off numbers

Concept 4: Factors and Multiples
Description:

1. Factor of a number is an exact divisor of that number
2. Every number is a multiple of any of its factors
3. 1 is a factor of all numbers
4. Every number is a factor of itself

Concept 5: Properties of Whole Numbers
Description:

1. Prime Number: All Numbers that have 2 factors, namely 1 and the number itself
2. Even Numbers: All Multiples of 2 are even numbers (OR) All numbers that end in 0, 2, 4, 6, 8 are even numbers
3. Odd Numbers: Numbers which are not multiples of 2 are odd numbers (OR) If the number is not an even number, then it is an odd number
4. Composite Number: Numbers with more than two factors are composite numbers

5. Twin Primes: Two consecutive odd prime numbers are called twin primes. Example: (3, 5), (5, 7), (11, 13), (17, 19)

6. Prime Triplet: Three consecutive odd prime numbers are called a prime triplet. The only prime triplet is (3, 5, 7)

7. Perfect Number: If the sum of all factors of the number is twice the number; then the number is a perfect number. Example: 6, 28

8. Co-primes: Two numbers are called co-primes, if they do not have a common factor other than 1. Example: 5, 9

   a. 1 is neither prime nor composite

   b. 2 is the smallest prime number

   c. 2 is the only even prime number

   d. All other even numbers are composite numbers

   e. Two prime numbers are always co-primes

   f. Two co-primes need not be prime numbers

Concept 6: Divisibility

1. By 2: Units place has 0, 2, 4, 6 or 8, then the number is divisible by 2

2. By 3: Sum of digits is divisible by 3, then the number is divisible by 3

3. By 4: Number formed by the units and tens place of a number is divisible by 4, then the number is divisible by 4

4. By 5: Units place has 0 or 5, then the number is divisible by 5

5. By 6: If the number is divisible by 2 and 3, then the number is divisible by 6

6. By 7:

   a. Split the number into Units and Rest

   b. New Number = (Rest - 2) × Units

   c. Check if this is divisible by 7; else split the number again

7. By 8: If the number formed by the last three digits of a number is divisible by 8

8. By 9: If the sum of the digits of the number is divisible by 9, then the number is divisible by 9

9. By 10: If the Units place is 0, then the number is divisible by 10

10. By 11:

   a. Method 1:

   i. Starting with Units place; determine the sum of odd digits and even digits

   ii. If the difference is 0 or a multiple of 11, then the number is divisible by 11

   b. Method 2:

   i. Group the number in pairs starting with the units

   ii. Add the numbers

   iii. If the sum is divisible by 11, then the number is divisible by 11

11. If a number is divisible by another, it must be divisible by the factors of the number. Therefore, every number that is divisible by 10 is also divisible by 5 and 2

12. If a number is divisible by two co-primes; it is divisible by their product

13. If a number is a factor of each of two numbers; then it is a factor of their sum and difference

Concept 7: Prime Factorization

Prime Factor: A factor of a number which is a prime number is called a Prime Factor

1. Prime Factorization: Expressing a number as a product of its prime factors

2. The largest common factor between two or more numbers is the Highest Common Factor or HCF

3. HCF of two co-primes is 1

4. HCF of a set of numbers is not greater than any of the numbers

5. The smallest multiple of the given set of numbers is called Least Common Multiple or LCM

6. For any two numbers, a and b, $a \times b = LCM \times HCF$

7. LCM of two co-primes is their product

8. LCM of a set of numbers is not less than any of the numbers

9. HCF is always a factor of the LCM

Concept 8: Whole Numbers and Integers

1. Natural Numbers: Counting Numbers are Natural Numbers

2. Whole Numbers: Natural Numbers together with 0 are Whole Numbers

3. Integers: Negative Numbers and Whole Numbers together form the set of Integers

4. Successor of a Whole Number: Add 1 to a whole number, we get the successor of the whole number

5. Predecessor of a Whole Number: Subtract 1 from a whole number, we get the predecessor of the whole number

6. 0 does not have a predecessor

7. Every other Whole Number has a predecessor and successor

8. 0 is not a Natural Number

9. 0 is the smallest Whole Number

10. 1 is the smallest Natural Number

11. 0 is lesser than every positive integer

12. 0 is greater than every negative integer

13. Absolute Value of an integer is the numerical value of the number regardless of its sign. Represented by $|\ |$. For example, $|-5| = 5$

Concept 9: Whole Numbers: Addition

1. Closure Property: If a and b are whole numbers, then $a+b$ is also a whole number

2. Commutative Law: If a and b are whole numbers, then $a+b = b+a$

3. Identity: If a is a whole number, then $a+0 = a = 0+a$

4. Associative Property: If a, b and c are whole numbers, then: $a+(b+c) = (a+b)+c = b+(c+a)$. The manner of association does not impact the sum

5. While adding a set of numbers, we write them down in columns and add them column wise. The order of addition of numbers in a column does not impact the final sum

Concept 10: Whole Numbers: Subtraction

1. Closure Property: If a and b are whole numbers, then $a - b$ is also a whole number if and only if $a \geq b$

2. Commutative Law: If a and b are whole numbers, then $a - b \neq b - a$. Subtraction is NOT commutative, if $a \neq b$

3. Identity: If a is a whole number, then: $a - 0 \neq 0 - a$. 0 is not a subtractive identity

4. Associative Property: If a, b and c are whole numbers, then: $a - (b - c) \neq (a - b) - c \neq b - (c - a)$. The manner of association does impact the difference. Associative property does not hold for subtraction

Concept 11: Whole Numbers: Multiplication

1. Closure Property: If a and b are whole numbers, then $a \times b$ is also a whole number

2. Commutative Law: If a and b are whole numbers, then: $a \times b = b \times a$

3. Identity: If a is a whole number, then $a \times 1 = a = 1 \times a$

4. Associative Property: If a, b and c are whole numbers, then: $a \times (b \times c) = (a \times b) \times c = b \times (c \times a)$. The manner of association does not impact the product

Concept 12: Whole Numbers: Division

1. $Dividend = (Divisor \times Quotient) + Remainder$

2. If a and b are non-zero whole numbers, then a/b is not necessarily a whole number

3. Division by 0 is not defined

4. If a is a whole number, then $\dfrac{0}{a} = 0$

5. If a is a whole number, then $\dfrac{a}{1} = a$

Concept 13: Properties of Integers: Addition

1. Closure Property: Sum of two or more integers is always an integer
2. Commutative law: If a, b are two integers, then: $a+b=b+a$
3. If a, b and c are integers, then: $a+(b+c)=(a+b)+c=b+(c+a)$. The manner of association does not impact the sum
4. If a is an integer, then $-a$ is the additive inverse of a, since: $a+(-a)=0$
5. If a is an integer, then $(a+1)$ is the successor; and $(a-1)$ is the predecessor of a

Concept 14: Properties of Integers: Subtraction

1. If a and b are integers, then: $a-b=a+(-b)$. Subtraction is the same as adding its additive inverse
2. Closure Property: If a and b are integers, then: $a-b$ is also an integer
3. If a is an integer, then: $a-0=a$
4. If a, b and c are integers such that $a>b>c$, then: $a-c>b-c$

Concept 15: Properties of Integers: Multiplication

1. Closure Property: Product of two integers is always an integer
2. Commutative Law: If a and b are integers, then: $a \times b = b \times a$
3. For every integer a: $a \times 1 = a$
4. Associative Law: If a, b, c are integers, then: $a \times (b \times c)=(a \times b) \times c = b \times (c \times a)$
5. Distributive Law: If a, b, c are integers, then: $a \times (b+c)=a \times b + a \times c$

Concept 16: Properties of Integers: Division

1. Closure Property: Does not hold; a/b is not necessarily an integer for all a and b
2. Division by 0 is not defined
3. If a is an integer, then: $a/1 = a$
4. If a is an integer, then: $a/a = 1$
5. If a is an integer, then: $0/a = 0$

Concept 17: Law of Signs

The laws of signs or the rules of signs, is enumerated below:
1. $+ \times + = +$
2. $\times + = -$
3. $+ \times - = -$
4. $\times - = +$
5. $+ / + = +$
6. $/ + = -$
7. $+ / - = -$
8. $/ - = +$

Concept 18: Fractions

1. If a and b are natural numbers, the number a/b is called a fraction
2. a/b is the same as dividing the whole into b **equal** parts, and considering a of them
3. In a fraction a/b, a is called numerator and b is called denominator
4. Fractions which represent the same part of the whole are called equivalent fractions. Therefore, $1/2 = 2/4 = 3/6 = 4/8$ are equivalent fractions
5. Multiplying the numerator and denominator by the same non-zero number m does not change the value of the fraction. Therefore: $a/b = (a \times m)/(b \times m)$, where m is a non-zero number

6. Dividing the numerator and denominator by the same non-zero number m does not change the value of the fraction. Therefore:

$a/b = \left(\dfrac{a}{m}\right) \Big/ \left(\dfrac{b}{m}\right)$, where m is a non-zero number

7. Two fractions, a/b and c/d are equivalent, if: $a \times d = b \times c$. This is the rule of cross products

8. Fractions with the **same** denominator are called **Like Fractions**

9. Fractions **without** the same denominator are called **Unlike Fractions**

10. Convert unlike fractions to like fractions

  a. Determine the LCM of the denominators of the unlike fractions

  b. Convert each fraction into an equivalent fraction with the LCM as its denominator

11. Simplest form or Lowest terms of a fraction: When the HCF of its numerator and denominator is 1

12. Proper Fraction: Fraction where numerator is less than denominator

13. Improper Fraction: Fraction where numerator is greater than the denominator

14. Mixed Fraction is simply the combination of an integer and a proper fraction

  a. If a is an integer and $\dfrac{b}{c}$ is a proper fraction, then $a\dfrac{b}{c}$ is called a mixed fraction.

  b. $a\dfrac{b}{c} = a + \dfrac{b}{c}$

  c. To convert $a\dfrac{b}{c}$ into an improper fraction

   i. $a\dfrac{b}{c} = \dfrac{a \times c + b}{c}$ where $\dfrac{a \times c + b}{c}$ is an improper fraction

  d. To convert an improper fraction $\dfrac{a}{b}$ into a mixed fraction

i. We divide a by b. Let q be the quotient and r be the remainder.

ii. $\dfrac{a}{b} = q\dfrac{r}{b}$

iii. Let us look at it the other way: $q\dfrac{r}{b} = \dfrac{q \times b + r}{b} = \dfrac{a}{b}$. And

$a = q \times b + r$

15. Comparison of fractions

   a. Among like fractions, the fraction with the larger numerator is larger

   b. Among fractions with the same numerator, the fraction with the larger denominator is smaller

   c. In general, we convert unlike fraction into equivalent fractions or like fractions and compare the numerators to determine the largest fraction

   d. If a/b and c/d are two fractions,

      i. We cross multiply and determine ad and bc

      ii. $\dfrac{a}{b} > \dfrac{c}{d}$ if $ad > bc$

      iii. $\dfrac{a}{b} < \dfrac{c}{d}$ if $ad < bc$

      iv. $\dfrac{a}{b} = \dfrac{c}{d}$ if $ad = bc$

16. Addition of Fractions

   a. Sum of like fractions $= \dfrac{\text{sum of numerators}}{\text{denominator}}$

   b. Sum of unlike fractions

      i. Convert unlike fractions into like fractions

      ii. Add the like fractions as shown above

17. Subtraction of Fractions

   a. Difference of like fractions $= \dfrac{\text{difference of numerators}}{\text{denominator}}$

b. Difference of unlike fractions

   i. Convert unlike fractions into like fractions

   ii. Subtract the like fractions as shown above

18. Addition and Subtraction of Mixed Fractions

   a. Convert mixed fractions to improper fractions

   b. Add / subtract the improper fractions using the techniques described above

19. Multiplication of Fractions

   a. Product of fractions $= \dfrac{\text{product of numerators}}{\text{product of denominators}}$

   b. We use the word "of" to indicate $\times$. Therefore a/b of c is the same as $a/b \times c = c \times a/b$

20. Reciprocal of a Fraction

   a. Two fractions are said to be reciprocal to each other, if their product is 1

   b. If $\dfrac{a}{b}$ is a non-zero fraction, then $\dfrac{b}{a}$ is the reciprocal fraction of $\dfrac{a}{b}$

21. Division of Fractions

   a. $\dfrac{a}{b} \Big/ \dfrac{c}{d} = \dfrac{a}{b} \times \left(\text{reciprocal of } \dfrac{c}{d} \right) = \dfrac{a}{b} \times \dfrac{d}{c}$

   b. It must be noted that both fractions have to be non-zero

   c. $a = \dfrac{1}{\left(\dfrac{1}{a} \right)}$

Concept 19: Brackets: Order of Simplification

Mnemonic: VBODMAS

   1. Resolve the brackets from the inner most pair

   2. Ensure that you account for the sign in front of the bracket

3. If we see a + sign in front of the bracket, every term inside the bracket is brought out as is, without any change in sign. The brackets are dropped.

4. If we see a − sign in front of the bracket, we change the sign of every term inside the bracket. The brackets are subsequently dropped.

5. The order of resolution or simplification of an expression are:

   a. Vinculum or bar

   b. Brackets [{( )}]

   c. Of

   d. Divide

   e. Multiplication

   f. Addition

   g. Subtraction

Concept 20: Decimals

1. Decimal Fraction : Fraction in which the denominators are 10, 100, 1000, … are known as decimal fractions

2. Tenths : A decimal fraction with 10 as the denominator

3. Hundredths: A decimal fraction with 100 as the denominator

4. Thousandths : A decimal fraction with 1000 as the denominator

5. Decimals : The numbers expressed in decimal from are called decimal numbers or simply decimals

6. Decimals have two parts

7. A whole-number part

8. A decimal part

9. These are separated by a dot

10. Decimal places : The number of digits contained in the decimal part of a decimal gives the number of decimal places. 2.85 has two decimal places

11. Like decimals : Decimals having the same number of decimal places are called Like decimals

15

12. Unlike decimals: Decimals having different number of decimal places are called unlike decimals.

13. Putting any number of zeros to the extreme right of a decimal does not change its value

14. Convert unlike decimals to like decimals: Add zeroes to the extreme right of the decimals so that the decimal places are equal.

15. Comparing decimals

    a. Convert the given decimals into like decimals

    b. Compare the whole part. Larger the whole part; larger the decimal

    c. If the whole parts are equal

        i. Compare the tenths digit. Larger the tenths digit; larger the decimal

        ii. If tenths digit are equal, go on to hundredths, thousandths and so on

16. Converting a decimal into a fraction

    a. Write the given decimal without decimal point as numerator

    b. Denominator is 1 followed by as many zero as there are decimal places

    c. Simplify the fraction

17. Converting a fraction into a decimal

    a. Divide numerator by the denominator till a non zero remainder is obtained

    b. Put a decimal point in the quotient and the dividend

    c. Put a zero to the right of decimal point in the dividend and right of the remainder

    d. Divide again just as we do in whole numbers

    e. Repeat steps 3 and 4 until the remainder is zero

18. Addition/Subtract in Decimals

    a. Convert the given decimals into like decimals

    b. Add/subtract in the case of whole numbers

19. Multiplication of a decimal by a decimal

a. Multiply the two decimals without the decimal point, just like whole numbers

b. Number of decimal points in the product is the sum of decimal points in the two decimals

c. Mark the decimal point

20. Division of Decimals

a. Convert decimals into like decimals

b. Eliminate the decimal point

c. Divide the two numbers just like two whole numbers

Concept 21: Percentages

1. Abbreviated as p.c. Symbol: %

2. Per cent means "per hundred" or "for every hundred"

3. By percentage, we mean "that many" hundredths

4. To convert a fraction into a percentage

5. $\dfrac{a}{b} = \left(\dfrac{a}{b} \times 100\right)\%$

6. To convert a percentage into a fraction

   a. $a\% = \dfrac{a}{100}$

7. Percentage as a ratio

   a. $a\% = \dfrac{a}{100} = a : 100$

8. Ratio as a percentage

   a. $a : b = \dfrac{a}{b} = \left(\dfrac{a}{b} \times 100\right)\%$

9. Percentage as a decimal

   a. $a\% = \dfrac{a}{100} =$; represent this as a decimal

10. Decimal as a percentage

    a. Convert the decimal into a fraction

b. Multiply the fraction by 100

c. Add a "%" symbol

11. Increase Percentage $= \left(\dfrac{\text{Increase}}{\text{Original Value}} \times 100 \right)\%$

12. Decrease Percentage $= \left(\dfrac{\text{Decrease}}{\text{Original Value}} \times 100 \right)\%$

Concept 22: Rational Numbers

1. **Fractions:** Numbers of the form a/b, where a and b are whole numbers; and $b \neq 0$

2. **Rational Numbers:** Numbers of the form a/b where a and b are integers; and $b \neq 0$

3. Observations

    a. Zero is a rational number since $0 = 0/1$; which is the quotient of two integers with a non-zero denominator

    b. Every natural number is a rational number

    c. Every integer is a rational number

    d. Every fraction is a rational number

4. **Positive Rational Number :** A rational number is said to be positive if its numeration and denominator are both positive or both negative

5. **Negative Rational Number :** A rational number is said to be negative, if its numerator and denominator have opposite sign

6. Properties of Rational Numbers

    a. If $\dfrac{p}{q}$ is a rational number; and m is a non-zero integer, then

    $\dfrac{p}{q} = \dfrac{p \times m}{q \times m}$. In other words, a rational number remains unchanged, if its numerator and denominator are multiplied by the same non-zero integer

b. If $\dfrac{p}{q}$ is a rational number, and m is a common divisor of p and q,

then $\dfrac{p}{q} = \dfrac{p/m}{q/m}$

c. The rational numbers that result from property (1) and (2) are called equivalent rational numbers

d. For every rational number x, exactly one of the following is true

   i. $x < 0$

   ii. $x = 0$

   iii. $x > 0$

e. For every pair of rational numbers x and y, exactly one of the following is true

   i. $x < y$

   ii. $x = y$

   iii. $x > y$

f. If x, y, z are three rational numbers, and $x > y$ and $y > z$, then $x > z$

7. Standard Form: A rational number $\dfrac{p}{q}$ is said to be in standard form, if q is positive, and p and q have no common divisor other than 1

8. Comparison of Rational Numbers

  a. Every positive rational number is greater than 0

  b. Every negative rational number is less than 0

  c. To compare two rational numbers

    i. Express the rational number with a positive denominator

    ii. Take the LCM of these denominators

    iii. Express the numbers as equivalent rational numbers with the LCM as its denominator

    iv. The number with the greater numerator is the greater

9. Addition of Rational Numbers

a. For any two rational numbers $\dfrac{p}{q}$ and $\dfrac{r}{q}$, we have:

$$\frac{p}{q} + \frac{r}{q} = \frac{p+r}{q}$$

b. If the denominators are the same, we simply add the numerators and divide the sum by the denominator

c. When the denominators are unequal

d. We take the LCM of the denominators

e. Express each rational number as an equivalent rational number with the LCM as the common denominator

f. Add the numerators and divined the sum by the common denominator, ie. the LCM

g. This can be represented as: $\dfrac{p}{q} + \dfrac{r}{s} = \dfrac{ps + rq}{qs}$

10. Subtraction of Rational Numbers

a. For any two rational numbers $\dfrac{p}{q}$ and $\dfrac{r}{s}$, we define:

$$\frac{p}{q} - \frac{r}{s} = \frac{ps - rq}{qs}$$

b. The rational number $\dfrac{r}{s}$ is the additive inverse of $-\dfrac{r}{s}$

c. Therefore, subtraction is simply addition of one rational number with the additive inverse of the other

d. $\dfrac{p}{q} - \left(-\dfrac{r}{s}\right) = \dfrac{p}{q} + \dfrac{r}{s}$

11. Multiplication of Rational Numbers

a. Product of two rational numbers $= \dfrac{\text{product of numerators}}{\text{product of denominators}}$

b. If $\dfrac{p}{q}$ and $\dfrac{r}{s}$ are two rational numbers, then: $\dfrac{p}{q} \times \dfrac{r}{s} = \dfrac{p \times r}{q \times s}$

c. If product of two rational numbers is equal to 1, then each of the rational numbers is called the reciprocal of the other

d. $\dfrac{p}{q}$ is the reciprocal of $\dfrac{q}{p}$

e. $\left(\dfrac{p}{q}\right)^{-1} = \dfrac{q}{p}$

f. Reciprocal for 0 does not exist

g. Reciprocal of 1 is 1

h. Reciprocal of −1 is −1

12. Division of Rational Numbers

a. If $\dfrac{p}{q}, \dfrac{r}{s}$ are rational numbers and $\dfrac{r}{s} \neq 0$, then

$$\dfrac{p}{q} \Big/ \dfrac{r}{s} = \dfrac{p}{q} \times \left(\dfrac{r}{s}\right)^{-1} = \dfrac{p}{q} \times \dfrac{s}{r} = \dfrac{p \times s}{q \times r}$$

Concept 23: Exponents

1. Repeated multiplication of a quantity a is represented using a short hand notation

2. $a \times a \times a \ldots (n \text{ times}) = a^n$. This is read out as "a to the power of n"

3. a is called the base

4. n is called the exponent

5. Therefore $a \times a = a^2$. This is also known as a-squared

6. $a \times a \times a = a^3$. This is also known as a-cubed

7. If $\dfrac{a}{b}$ is a rational number, then

a. $\left(\dfrac{a}{b}\right)^n = \dfrac{a^n}{b^n}$

b. The reciprocal of $\left(\dfrac{a}{b}\right)^n$ is $\left(\dfrac{b}{a}\right)^n$

c. Therefore, reciprocal of $\dfrac{a^n}{b^n}$ is $\dfrac{b^n}{a^n}$

d. $\left(\dfrac{a}{b}\right)^m \times \left(\dfrac{a}{b}\right)^n = \dfrac{a^m}{b^m} \times \dfrac{a^n}{b^n} = \dfrac{a^{m+n}}{b^{m+n}} = \left(\dfrac{a}{b}\right)^{m+n}$

e. if $m > n$, then $\left(\dfrac{a}{b}\right)^m \Big/ \left(\dfrac{a}{b}\right)^n = \dfrac{a^m}{b^m} \Big/ \dfrac{a^n}{b^n} = \dfrac{a^{m-n}}{b^{m-n}} = \left(\dfrac{a}{b}\right)^{m-n}$

f. if $m < n$, then $\left(\dfrac{a}{b}\right)^m \Big/ \left(\dfrac{a}{b}\right)^n = \dfrac{a^m}{b^m} \Big/ \dfrac{a^n}{b^n} = \dfrac{1}{a^{n-m}/b^{n-m}} = \left(\dfrac{a}{b}\right)^{m-n}$

g. $\left(\dfrac{a}{b}\right)^0 = 1$

h. $\left(\dfrac{a}{b}\right)^{-1} = \dfrac{b}{a}$

i. $a^{-1} = \dfrac{1}{a}$

j. $\left(\dfrac{a}{b}\right)^{m-n} = \dfrac{1}{a^{n-m}/b^{n-m}}$

Concept 24: Ratio

1. Ratio lets us compare **two similar** quantities

2. Comparison by difference will determine if one quantity is lesser than, equal to or greater than the other. We can say x is greater than y if $(x - y) > 0$

3. Comparison by division is another method. Here we compare two quantities by determining $r = \dfrac{x}{y}$, where r is called the ratio of x to y

4. Ratio has no units; because the two quantities are similar. Ratio of two quantities is simply a fraction that one quantity is of the other

5. For two non-zero quantities x and y, the ratio x to y is a ratio $\frac{x}{y}$. It is written as $x : y$

6. x is known as antecedent and y is known as consequent

7. A ratio can be expressed as a fraction $\frac{x}{y}$

8. Therefore, $x : y = mx : my$ for $m \neq 0$

9. Ratio $a : b$ is greater than $x : y$ if $ay > bx$. Ratio $a : b$ is less than $x : y$ if $ay < bx$. Ratio $a : b$ is equal to $x : y$ if $ay = bx$

10. When the two terms of a ratio have no common factor other than 1, we say that the ratio is in its simplest form

11. We usually represent a ratio in its simplest form

12. Comparison of two ratios is meaningful if and only if the ratios are of the same kind

13. Comparison of Ratios

   a. Express the ratios as fractions

   b. Convert the fractions into equivalent fractions

   c. Compare

   d. Simple Method

      i. Let $a : b$ and $c : d$ be two ratios to be compared

      ii. Therefore, we can represent them as fractions: $\frac{a}{b}$ and $\frac{c}{d}$

      iii. The equivalent fractions are $\frac{ad}{bd}$ and $\frac{bc}{bd}$

      iv. If $a \times d > c \times b$, then $a : b > c : d$

      v. If $a \times d < c \times b$, then $a : b < c : d$

      vi. If $a \times d = c \times b$, then $a : b = c : d$

Concept 25: Unitary Method

Here is a common problem solving scenario

1. We are typically given the value x for y things

2. We are expected to determine the value for z things

3. We can use the unitary method for determining this

Unitary Method is as follows

1. Given: Value of y things is x

2. Find the value of 1 thing: y / x

3. Therefore, value of z things is value of 1 thing $\times$ z things. This is the same as $z \times y / x$

4. This is a simple outline of the unitary method

5. Summary: Find the value of a one (unit) thing; multiply this by the number of things you desire. This gives you the final answer

Concept 26: Proportion

1. Four numbers a, b, c, d are said to be in proportion if $\dfrac{a}{b} = \dfrac{c}{d}$

2. Expressed as $a : b = c : d$ or $a : b :: c : d$

3. a, d are called extreme terms or extremes

4. b, c are called middle terms or means

5. If product of the means is equal to the product of the extremes, $b \times c = a \times d$, then $a : b :: c : d$ (OR) we say that a, b, c, d are in proportion

6. If $a : b :: c : d$, then

   a. $a : c :: b : d$

   b. $b : a :: d : c$

   c. $c : a :: h : b$

7. If $a : b :: b : c$ then $b^2 = ac, b = \sqrt{ac}$; b is called the geometric mean of a and c

8. Direct Proportion or Variation

   a. Two quantities a and b, are said to vary directly, if the ratio $a : b$ remains constant

   b. For example,

    i. Total cost of articles increases when the number of articles increases. In other words, it costs more to get more

    ii. Total work done increases when the number of men at work increases. In other words, when you employ more men, you get more work done

  c. Two quantities are said to vary directly if the increase or decrease in one quantity causes a proportional increase or decrease in the other

  d. $\dfrac{a}{b} = k$ or $a = k \times b$, where k is a constant

9. Inverse Proportion or Variation

  a. Two quantities a and b, are said to vary inversely, if the product $a \times b$ remains constant

  b. For example

    i. To cover a certain distance, time taken decreases when the speed of the car increases

    ii. To complete a certain amount of work, time taken decreases when the number of men employed increases

  c. Two quantities are said to vary inversely if an increase or decrease in one quantity causes a proportional decrease or increase in the other

  d. $a \times b = k$ or $a = \dfrac{k}{b}$, where k is a constant.

Concepts in
Elementary Algebra

Concept 1

Algebra is similar to Arithmetic. In both cases, we manipulate quantities. Arithmetic deals with manipulation of numbers. Algebra is slightly different. In Algebra, we deal with a greater generality. We employ letters or symbols to denote quantities on which we perform the mathematical operations or manipulations.

The term 'manipulation' means the same in both the worlds. A set of inputs or information about a problem is given. We will derive the result using the inputs. Thus, from a problem-solving standpoint, both Arithmetic and Algebra are very similar.

Each number represents a unique point in the number line. An algebraic symbol stands for one or more numerical values. In several cases, we can have a symbol that can stand for an infinitely large number of numeric values. It is possible for us to apply the mathematical operations without assigning any specific value to a symbol. This may confuse the beginner. This will become clearer as we make progress.

The meanings of the mathematical operations such as addition, subtraction, multiplication and division remain the same.

Example: Let us look at the following statement. "Jack is 5 years older than Jill. Jill's age is 15."

Let Jill's age be x years. Then, Jack's age will be $(x + 5)$ years.

In the above formulation, the unknown quantity "x" is a symbol used to represent the age of Jill. It could be any number of years. We know that Jack's age is simply "Jill's age plus 5 years". This reality is represented as $(x + 5)$ years.

Here x is a variable. It can assume any numerical value as determined by the final solution to the problem. x is also a symbol which stands for Jill's age. This is a fundamental concept, is all-pervasive, in Algebra.

Concept 2

An algebraic term is simply a variable with or without a coefficient. . Therefore, x is a variable, a literal or a symbol (we will use these terms interchangeably). $7x, 8x$ are examples of algebraic terms. When the coefficient is absent, we assume that it is equal to one. Therefore, the term x is the same as $1x$.

An algebraic expression is simply a collection of algebraic terms connected by mathematical operations such as $+, -, \times$ and $\div$. There are other mathematical operations as well. We will consider those in the subsequent sections, when appropriate.

For example, $7a + 6b + 8c - d - 16z$ is an algebraic expression that has 5 terms in it.

An algebraic term $7x$ is the same as $+7x$. In other words, we assume the sign to be $+$, whenever the sign is not explicitly specified. This is similar to arithmetic concept of numbers. 8 is the same as $+8$. And, $197 = +197$.

Concept 3

A simple expression consists of one term. $8x, 9y, 16a$ are examples of a simple expression. A compound expression consists of several terms. $8x + 9y - 16a$ is a compound expression. A binomial is a compound expression with two terms. A trinomial is a compound expression with three terms. In addition, a multinomial is a compound expression with more than three terms.

For example, $3x + 2y$ is a binomial expression. $6a + 7b - 3c$ is a trinomial and $8x - 4y - 23z + 16a - 2d$ is a multinomial.

Concept 4

Just like in arithmetic, when two or more variables, terms or expressions are multiplied together, we get a product. However, there is one big difference in representation.

The product of two variables, a and b, is ab. This is the same as $a \times b$.

If $a = 9$ and $b = 4$, $ab = a \times b = 9 \times 4 = 36$.

In arithmetic, 4×5 is 20 and not 45. We cannot drop the product sign in arithmetic. In fact, 45 is $4 \times 10 + 5$.

Concept 5

In arithmetic, $24 = 2 \times 3 \times 4$. Therefore, two, three and four are factors of 24. Similarly, in an algebraic term $7abc$ has four factors. These are seven, a, b and c. It would be important for us to recall that $7abc = 7 \times a \times b \times c$.

Concept 6

The numeric part of an algebraic term is usually called the coefficient of the term. Therefore, 6 is the coefficient of ab in $6ab$. In the algebraic term abc, we can refer to a as the literal coefficient of bc. Therefore, there are situations where a coefficient is not necessarily a numerical quantity. In general, when we simply talk about coefficients, we refer to the numerical part of a term.

The term a is the same as $1a$. So, when the coefficient of a term is 1, we usually ignore this and simply write the variables involved. Therefore, $abc = 1abc$. Or, $1xyz = xyz$.

Concept 7

If we consider a quantity or a variable x, and repeated multiply the variable by itself, we get a product. This product is called power of the variable or quantity. Therefore,

$a \times a$ is written as a^2. This is the second power of a. This is also read out as "a to the power of 2". This is also called a-squared.

Similarly, $a \times a \times a = a^3$. This is the third power of a. This is is read out as "a to the power of 3". This is also called a-cubed.

And, $a \times a \times a \times a \cdots n$ times $= a^n$. This is the nth power of a. This is read out as "a to the power of n".

In these example, a is called the base and the number which represents the power is called the index or exponent.

The quantity a is the same as $a1$, which is the first power of a. Therefore,

$a = 1a = a^1 = 1a^1$. They all have the same meaning and the 1 is omitted in each of these cases.

By definition, any base to the power of zero is 1. Zero to the power of any anything is zero. Therefore,

$a^0 = 1$ and $0^n = 0$

Every power of 1 is equal to 1. Therefore $1^2 = 1^3 = 1^9 = 1^n$.

Concept 8

Conceptually, a coefficient and exponent (or index) are very different things. It is critical to send a moment to understand the difference - since both seem to fit the definition of being the numerical part of a term ☺

$4a$ means 4 times a. $3b$ means 3 times b.

a^4 means $a \times a \times a \times a$ or a multiplied by itself four times. Similarly, b^3 means $b \times b \times b$.

If $a = 3$ and $b = 4$, then $4a = 4 \times 3 = 12$; $a^4 = 3 \times 3 \times 3 \times 3 = 81$. Similarly, $3b = 3 \times 4 = 12$; and $b^3 = 4 \times 4 \times 4 = 64$.

Concept 9

Like in the arithmetic world, multiplication is commutative in the algebraic world as well. In other words, just like $4 \times 5 = 5 \times 4$ in the arithmetic world; $ab = ba$ in the algebraic world. The order in which we multiply the constituent quantities in a term has no impact on the final product. xyz, yxz and yzx are all equal to one another - they produce the same final product.

Concept 10

$aaaabbbcccc$ is the same as $a^4 b^3 c^5$. Similarly, $16x^2 y^3 z^6$ is the same as $16xx\ yyy\ zzzzzz$. In other words, you can multiply several quantities, each raised to some power; the basic notation of multiplication applies with no exception.

Therefore, if $a = 3; b = 4; c = 5$, $16a^2bc^3 = 16 \times 3 \times 3 \times 4 \times 5 \times 5 \times 5 = 72000$.

Concept 11

If one of the factors of a product is zero, then the product is always zero, irrespective of the number of other non-zero factors in the product. This can be simply stated as "anything multiplied by zero gives a zero".

Therefore, we can also conclude that every power of zero is zero. $0 = 0^1 = 0^9 = 0^n$

Concept 12

There is a distinction between sum of two terms and product of two terms. The sum of two terms a and b is $a + b$; whereas the product of the two terms is ab. We cannot ignore any other mathematical symbol. The only symbol that is ignored in algebra is the multiplication symbol.

Concept 13

In an arithmetic expression, quantities in an expression prefixed with $a +$ sign are additive terms and those prefixed with $a -$ sign are subtractive terms.

In algebraic expression, the same concept of additive and subtractive terms exists. Therefore, in the expression $3a - 2b - 3c$; $3a$ is an additive term and $-2b$ and $-3c$ are additive terms.

Concept 14

The sum of all additive terms in an arithmetic expression is always greater than the sum of all subtractive terms. In an arithmetic expression $4 + 7 - 5 - 9$; the sum of $+4$ and $+5$ is always greater than the sum of -5 and -9.

The same statement cannot be made in the case of an algebraic expression. It is perfectly possible for the sum of subtractive terms to be larger than the additive terms. In the above expression $3a - 2b - 3c$, it is possible for $-2b + -3c$ to be greater than $3a$. In fact a subtractive term may also have a meaning even when it stands by itself.

Concept 15

If you started off with $100; made a profit of $70 and then a loss of $30; then you will be left with $100 + $70 - $30 = $140. In other words, you would have made a profit of $40.

The algebraic statement for this is $140 - $100 = $40

If you started off with $100; made a loss of $70 and then a profit of $30; then you will be left with $100 - $70 + $30 = $60. In other words, you would have made a loss of $40.

The algebraic statement for this is $60 - $100 = -$40

The negative sign indicates net loss.

Concept 16

A subtractive quantity is always opposite of an additive quantity. This is true of arithmetic and algebra. $+4$ and -4 are equidistant from 0, but on either sides of 0. In that sense, we can look at subtraction as the reverse operation, or opposite of addition.

Concept 17

Sets of algebraic terms that differ only in the coefficients are called like terms. All terms that are not like terms are called unlike terms. $3a, -7a$ and $12a$ are like terms. $3a, 3a^2$ and $-3ab$ are unlike terms.

Concept 18

Addition of algebraic expression follows the following rules:

1. Collect all like terms together
2. Add all the numerical values of coefficients of all the additive like terms of each set of like terms. Prefix the resultant sum with $a+$ sign.
3. Add all numerical values of coefficients of all the subtractive like terms of each set of like terms. Prefix the resultant sum with $a-$ sign.
4. Add the two resultant sums of the coefficients to get the final sum of each of the sets of like terms.

For example, while adding $12a - 26a + 13a + 6a - 2a$, the sum of additive like terms is $12 + 13 + 6 = 31$. Prefix this with $a+$ sign. We get $+31$. The sum of additive like terms is $26 + 2 = 28$. Prefix this with $a-$ sign. We get -28. Add these two together. We get $31a - 28a = 3a$.

Concept 19

The sum of two like terms with the same numerical coefficient; but different in sign is always zero. Therefore $3a - 3a = 0$; $5a - 5a = 0$ and $16xy - 16xy = 0$.

Concept 20

The algebraic sum is independent of the order in which the terms are added. In other words, $x++y = y+x$, and $x-y = -y+x$. $x+y-z = -z+y+x = y-z+x$.

Concept 21

We use brackets to group sets of terms together. It is also useful to unambiguously indicate what we are planning to do. When we say $x+(y+z)$ indicates that we wish to add y and z. Then add the resultant sum to x. $x+(y-z)$ indicates that we want to add the difference of y and z, to x.

$$x+(y-z) = x+y-z$$

Concept 22

When we come across $a+$ sign in front of the brackets; we can safely drop the brackets, with no change to signs of any of the terms in the expression.

$$x+y-z+(a+b-c-d) = x+y-z+a+b-c-d$$

Concept 23

When we come across $a-$ sign in front of the brackets; we can drop the brackets, and change the sign of each of the terms inside the brackets.

$$x+y-z-(a+b-c-d) = x+y-z-a-b+c+d$$

Concept 24

From the above two concepts, the following observations can be made.
1. $a+(+b) = a+b$
2. $a+(-b) = a-b$
3. $a-(-b) = a+b$
4. $a-(+b) = a-b$

Concept 25

Multiplication is repeated addition. Therefore,

$xy = x + x + x \cdots y$ times

$= yx$

$= y + y + y \cdots x$ times

Concept 26

The order of multiplication has no impact on the resultant product. This is also known as the associative law of multiplication.

$xyz = x(yz) = (xy)z = y(xz)$

Concept 27

Coming to the concept of repeated multiplication, we can make the following observations:

$a^n = a \times a \times a \times a \cdots n$ times

$a^m = a \times a \times a \times a \cdots m$ times

$a^n \times a^m = (a \times a \times a \times a \cdots n$ times$) \times (a \times a \times a \times a \cdots m$ times$)$

$= (a \times a \times a \times a \cdots m + n \text{times}) \ x = a^{(m+n)})$

In these expressions, a is called the base of the term and m and n are called the indices.

Concept 28

Similarly,

$a^n = a \times a \times a \times a \cdots n$ times

$a^m = a \times a \times a \times a \cdots m$ times

$a^n \div a^m = (a \times a \times a \times a \cdots n$ times$) \div (a \times a \times a \times a \cdots m$ times$)$

$= (a \times a \times a \times a \cdots m - n$ times$) = a^{(m-n)}$

Concept 29

As a consequence of the aforementioned concepts, we can come to the following definitions:

1. $a^0 = 1$, by definition
2. $1^n = 1$
3. $0^n = 0$

4. $0^0 = 0$

Concept 30

Let is now consider the expression $m(x + y)$.

$m(x + y) = (x + y)m$
$= ((x + y) + (x + y) + (x + y) + \cdots m \text{ times })$
$= ((x + x + x \cdots m \text{ times}) + (y + y + y \cdots m \text{ times}))$
$= mx + my$

Similarly, $m(x - y) = mx - my$

Therefore, $m(x - y - z) = mx - my - mz$

Concept 31

$m(a + b) = ma + mb$

Let $m = (c + d)$

$(c + d)(a + b) = (a + b)(c + d) = c(a + b) + d(a + b)$
$= ac + bc + ad + bd$

Similarly, $(a - b)(c - d) = a(c - d) - b(c - d)$
$= ac - ad - (bc - bd)$

Now, we remove the brackets. We have $a -$ sign in front of the bracket. Therefore sign of each of the term inside the brackets change.

$= ac - ad - bc + bd$

Concept 32

The rule of sign can be summarized from the previous concept. Let us recall the findings.

1. $+a \times +c = +ac$
2. $+a \times -d = -ad$
3. $-b \times +c = -bc$
4. $-b \times -d = +bd$

The rule of signs emerges from the above observations.

1. $+ \times + = +$
2. $+ \times - = -$

3. $- \times + = -$
4. $- \times - = +$

Concept 33

We can now write down the product of compound expressions based on the concepts we have covered so far.

1. $(x+a)(x+b) = x^2 + xb + ax + ab = x^2 + x(a+b) + ab$
2. $(x-a)(x+b) = x^2 + xb - ax - ab = x^2 + x(-a+b) - ab$
3. $(x+a)(x-b) = x^2 - xb + ax - ab = x^2 + x(a-b) - ab$
4. $(x-a)(x-b) = x^2 - xb - ax + b = x^2 - x(a+b) + ab$

Concept 34

The conclusions drawn in the previous concept lead us to the technique for writing down the products based on inspection.

When the compound expression is of the form $(x+a)(x+b)$, we can note the following:

1. The product has three terms.
2. The x^2 term has a coefficient of 1
3. The x term has a coefficient of $(a+b)$ We need to ensure that we take the signs of a and b into account as well.
4. The constant term is simply the product of a and b. Again, we need to take the signs of a and b into account.

Concept 35

Division is the only operation when produces two results. While the primary purpose of division is to determine the quotient, the operation also leaves behind a remainder; which is sometimes positive non-zero; but always less than the divisor. This means remainder can never be negative or a number or expression greater than the divisor.

Concept 36

This can be represented as:- dividend $\div$ divisor $=$ quotient; remainder

Therefore,

quotient $\times$ divisor $+$ remainder $=$ dividend

When remainder is equal to zero, this expression reduces to

quotient $\times$ divisor $=$ dividend

Concept 37

Each multiplication fact yields two division facts. Therefore, the rule of sign holds for division.

1. $+a \div +b = +c$
2. $+a \div -b = -c$
3. $-a \div +b = -c$
4. $-a \div -b = +c$

Just like the multiplication, clearly, like signs produce $+$ and unlike signs produce $-$.

Concept 38

We have already seen $a^m \div a^n = a^{(m-n)}$

Concept 39

Dividing each term of the expression by the divisor term completes division of an expression by a term. We retain or change the signs of the terms in the expression based on the rule of signs.

Therefore, $(ax^2 + bx + c) \div m = \dfrac{ax^2}{m} + \dfrac{bx}{m} + \dfrac{c}{m}$, where is may be any algebraic term.

Concept 40

We now address the method for dividing one algebraic expression by another algebraic expression.

1. Step 1: Arrange the dividend and divisor in descending powers of some literal in the expression. In most expressions, this literal is x; but this is not the rule. We will assume that this is x for purposes of explaining this rule. We also insert missing powers of x in the dividend and divisor with a 0 coefficient.
2. Step 2: Divide the term with the highest power of x in the dividend, with the term with the highest power of x in the divisor. We get the partial quotient.

3. Step 3: Multiply the partial quotient with the divisor and subtract this expression from the dividend.
4. Step 4: We then bring down the necessary terms of the dividend.
5. Step 5: We repeat step 2 to step 5 until all terms of dividends are brought down.

Concept 41

We know that $a \times a = a^2$. a is called the square root of a^2. This is also denoted as **root a.**

Concept 42

Similarly, if a is the cube root of a^3; a is the fourth root of a^b and a is the fifth root of a^5. Generalizing this observation, we have a is the nth root of a^n.

Concept 43

The number of literals in an algebraic term, excluding the numerical coefficient, is the degree of the term. Each literal constitutes a dimension of the term. Therefore, $4xyz$ has three dimensions and is of third degree.

Concept 44

The highest dimensions in an expression are also known as the degree of an expression. For example, $ax^2 + bx + c$, is an expression in second degree in x.

Concept 45

When all terms in a compound expression are of the same dimensions, we call the expression a homogenous expression. For example, $a^3 + b^3 + 3a^2b + 3ab^2$ is a homogenous compound expression.

The product of two or more homogenous expression is always homogenous. Similarly, when we divide a homogenous expression by another homogenous expression, the quotient is also homogenous.

Concept 46

This concludes our review of the concepts of basic operations like addition, subtraction, multiplication and division of algebraic terms and expressions. We have uses arithmetic operations and principles as the golden reference. We have compared the corresponding algebraic realities against this gold standard.

Concept 47

We now turn our attention to the important concept of equation. This is a central notion in algebra. When we assign the resultant value of an algebraic term or an expression to another algebraic term or an expression or a number; we refer to this assignment of equality as an equation.

Therefore, we can enumerate a few examples of equations using this definition.

1. $2x + 5 = 9y$
2. $7y + 2x = 5$

Concept 48

Each equation has three parts. An "equal" sign [the middle part] that separates the expression to its left [Left hand side or LHS] and the expression to its right [Right hand side or RHS] constitutes an equation. The most generic form of an equation therefore is

LHS expression = RHS expression.

Concept 49

An equation that is valid for all values of "x", the unknown is called an identity. For example,

$(x + a)^2 = x^2 + 2ax + a^2$ is an identity because this is valid for all values of x and a. x and a are the unknowns in this equation.

Concept 50

Certain equations are conditionally true. This means that these equations are true only for certain values of x or their unknowns. These are called conditional equations.

1. $9x = 27$ is valid only if $x = 3$ and for no other value of x.

2. $16x = 256$ is valid for $x = 16$ and for no other value of x.

3. $x^2 = 81$ is true when $x = 9$ or $x = -9$ and for no other value of x.

In these examples, the **solutions** simply are value of x, which satisfies the equation. In $3x = 9$; $x = 3$ is a **solution**. 3 is said to **satisfy** the equation. x is the **unknown quantity** in the equation. The process of determining the value of the **unknown quantity** that **satisfies** the equation is called **solving the equation**.

Concept 51

An equation that involves an unknown quantity in the first degree is also known as simple equation. The solution to a simple equation can be determined by applying the following axioms to the equations:

1. Adding the same quantity to both sides of the equation does not affect the equality.
2. Subtracting the same quantity to both sides of the equation does not affect the equality.
3. Multiplying the same quantity to both sides of the equation does not affect the equality.
4. Dividing the same non-zero quantity to both sides of the equation does not affect the equality.

Concept 52

Principle of transposition follows from the axioms mentioned in the preceding section. We will derive these from first principles.

Consider the equation,

$$x = a$$

Subtract a from both sides of the equation; the equality is not impacted.

$$x - a = a - a$$

$$x - a = 0$$

This means a can be transposed from RHS to the LHS with a change in sign. Generalizing the observation, we can conclude that **any term can be transposed from one side o the equation to the other side by changing its sign**.

Consider the equation, $mx = a$

Dividing both sides by m makes no change in the equality.

$$mx \div m = a \div m$$

$$x = \frac{a}{m}$$

Thus **a term, which is a factor on one side, becomes the divisor on the other side. Similarly a divisor on one side of the equation becomes a factor on the other side** after transposition operation.

Concept 53

The process for solving simple equations is straightforward.
1. We transpose all terms involving the unknown quantity to the left hand side of the equation.
2. We transpose all other terms to the right hand side of the equation.
3. We divide both sides of the equation by the coefficient of the unknown quantity to get the value of the unknown quantity.

Concept 54

Verification of the solution is an important part of the problem solving process. Verification ensures accuracy of the solution. It is also recommended that you substitute the solution in the original equation and ensure that the equation is satisfied.

Concept 55

Armed with the fundamentals for manipulating algebraic expressions, we can now look at concepts of symbolic expressions. Algebra may be treated as a mathematical language used for representing real world problems that we face. These problems typically appear in English or your native language. Algebra ensures that the problems are represented by means of symbols, expressions and relationships, which can be looked at as a whole and solved.

Concept 56

The common factor, which is of the highest dimension, which divides a set of expressions without leaving a remainder, is called the highest common factor. This is also referred to as HCF [highest common factor] or GCF [greatest common factor] or GCM [greatest common measure]. We will use HCF to denote this.

Concept 57

In case of simple expressions, we can determine the HCF by inspection.

1. Step 1: Determine the HCF of the numerical coefficients of the expressions or algebraic terms.
2. Step 2: For each literal check the highest power which will divide each of the expression without leaving a remainder. This is also the least power of that literal across all the expressions.
3. The HCF is simply the product of step 1 and step 2.

For example, HCF of $6a^3b^6$, $9a^2b^9$ and $12a^6b^7$ is $3a^3b^6$.

Concept 58

The least common multiple is simply the expression with the lowest dimension, which is divisible by each of the expressions without leaving a remainder. This is abbreviated as LCM.

Concept 59

We can determine the LCM of simple expressions by inspection.

1. Step 1: Determine the LCM of the numerical coefficients of the expressions or algebraic terms.
2. Step 2: For each literal check the lowest power which is divisible by each of the expression without leaving a remainder. This is also the highest power of that literal across all the expressions.
3. The LCM is simply the product of step 1 and step 2.

For example, LCM of $6a^3b^6$, $9a^2b^9$ and $12a^6b^7$ is $36a^6b^9$.

Concept 60

The concept of fractions is the same in arithmetic and algebra. Let us commence our discussion of fractions with the definition of fractions. If a whole is split into b equal parts, and we consider a of those parts; we are talking of $\dfrac{a}{b}$ of the whole.

In arithmetic, the whole is usually treated as 1. The whole in the world of algebra is an unknown quantity x. The resultant fraction is $\frac{a}{b}$ of x refers to a parts of b equal parts that x is divided into.

Concept 61

In order to reduce the fraction to its lowest terms, we eliminate the common factors from the numerator and denominator. The fraction that remains behind after this process of canceling common factors is the lowest terms of the fractions.

Concept 62

The product of two algebraic fractions is similar to the case in arithmetic. We divide the product of numerators by the product of denominators. We complete the operation by converting the resultant fraction by lowest terms.

Concept 63

The reciprocal of a fraction $\frac{a}{b}$ is $\frac{b}{a}$. The notion of a reciprocal is an important to the concept of division in fractions.

Concept 64

Two divide one fraction by another fraction, we multiply the first fraction by the reciprocal of the second fraction. In other words,

$$\frac{a}{b} \div \frac{c}{d} = \frac{a}{b} \times \frac{d}{c}$$

Concept 65

The concept of equivalent fractions is central to the operations of addition and subtraction of fractions. Let us now take a look at the concept of equivalence. This is similar to the world of arithmetic.
1. We take the LCM of the denominators of the fractions.
2. We multiply the numerators by the quotient that we get by dividing the LCM by the denominators.
3. The resultant fractions are equivalent fractions.
Let us consider the following example.

$$\frac{a}{3xy}, \frac{b}{6xyz}, \frac{c}{2yz}$$

The LCM of denominators $-3xy, 6xyz$ and $2yz$ is $6xyz$.

Therefore $\dfrac{a}{3xy} = a \times \dfrac{2z}{6xyz} = \dfrac{2az}{6xyz}$

Similarly, $\dfrac{b}{6xyz} = \dfrac{b}{6xyz}$ [no change required since denominator = LCM]

And, $\dfrac{c}{2yz} = \dfrac{3cx}{6xyz}$.

The equivalent fractions of $\dfrac{a}{3xy}, \dfrac{b}{6xyz}, \dfrac{c}{2yz}$ is $\dfrac{2az}{6xyz}, \dfrac{b}{6xyz}, \dfrac{2az}{6xyz}$ respectively.

Concept 66

In order to add or subtract fractions,
1. Step 1: Convert the fractions to its lowest terms.
2. Step 2: Convert the fractions to equivalent fractions.
3. Step 3: Add or subtract the numerators as required
4. Step 4: The common denominator of the resultant fraction is the common denominators of the equivalent fractions.
5. Step 5: Reduce the fraction to lowest terms if required.

Concept 67

Involution is simply the process of multiplying an expression or term with itself to find the second, third or nth power of the expression or term.

Concept 68

Let us look at the following set of statements.

$a^m = a \times a \times a \cdots m$ times

$(a^m)^n = (a \times a \times a \cdots m\,\text{times}) \times (a \times a \times a \cdots m\,\text{times})$
$\times (a \times a \times a \cdots m\,\text{times}) \cdots n$ times

$$= (a \times a \times a \cdots m \times n \, \text{times})$$

$$= a^{mn}$$

Concept 69

1. Every even power of an expression is always positive.
2. Every odd power of an expression retains the sign of the original expression

Concept 70

To find out the nth power of an expression:
1. Raise the numeric coefficient to the required power
2. Prefix the same with appropriate sign based on Concept 69
3. Raise the expression to the required power using Concept 68
4. The product of the previous three steps is the required power of the algebraic expression

Concept 71

The few common expressions:
1. $(a+b)^2 = a^2 + 2ab + b^2$
2. $(a-b)^2 = a^2 - 2ab + b^2$
3. $(a+b+c)^2 = a^2 + b^2 + c^2 + 2ab + 2bc + 2ac$

The general approach to finding the square of a multinomial expression is as follows.

    a.   First create an expression with the sum of squares of each of the terms in the multinomial. Squares are always positive no matter the sign associated with the term is.

    b.   Now add twice the product of each of the terms with every other term. Ensure that the sign of each term is accounted for while taking the product of two terms.

We can also look at the expression for cubes of sum of terms.
1. $(a+b)^3 = (a+b)(a+b)^2 =$
   $(a+b)(a^2 + 2ab + b^2) = a^3 + b^3 + 3a^2b + 3ab^2$

2. $(a-b)^3 = (a-b)(a-b)^2 =$
   $(a-b)(a^2+b^2-2ab) = a^3 - b^3 - 3a^2b + 3ab^2$

Concept 72

Evolution is the opposite of involution. The operation of finding the nth power of a term is involution. The operation of finding the nth root of a number is evolution. Evolution of a term x, is another term y, such that $x = y^n$. Therefore the nth root of x is y.

Concept 73

The rule of signs for evolution can be summarized as below:
1. An even root of any positive term or quantity can be either positive or negative. This is because product of two positive terms or two negative terms is always positive.
2. A negative term or quantity cannot have an even root.
3. Every odd root of a term or quantity retains the sign of the term or quantity. This is because, the product of odd number of -1s is -1.

Concept 74

The nth root of an expression is the same as raising the expression to the power of $\dfrac{1}{n}$. Therefore, the nth root of $a^m = (a^m)^{\frac{1}{n}} = a^{\left(\frac{m}{n}\right)}$.
1. Therefore, the square root of a term or expression is the same as raising the term or expression to the power of $\dfrac{1}{2}$.
2. Similarly, the cube root of a term or expression is the same as raising the term or expression to the power of $\dfrac{1}{3}$.

Concept 75

The method of determining the root of a simple expression is as follows:
1. Determine the root of the numerical coefficient.

2. Divide the exponent of each of every factor in the expression by the index of the root.

Concept 76

The square root of common expressions:
1. $(a+b)^2 = a^2 + 2ab + b^2$, therefore the square root of $a^2 + 2ab + b^2$ is $(a+b)$.
2. $(a-b)^2 = a^2 - 2ab + b^2$; therefore the square root of $a^2 - 2ab + b^2$ is $(a-b)$.
3. $(a+b+c)^2 = a^2 + b^2 + c^2 + 2ab + 2bc + 2ac$; therefore the square root of $a^2 + b^2 + c^2 + 2ab + 2bc + 2ac$ is $(a+b+c)$

Similarly, we can make the following conclusions.
1. The cube root of $a^3 + b^3 3ab^2$ is $(a+b)$.
2. The cube root of $a^3 - b^3 - 3a^2b + 3ab^2$ is $(a-b)$

Concept 77

When an algebraic term or an expression is a product of two or more factors, the process of determining these factors is called factorization or resolution into factors.

Concept 78

For any algebraic term, the coefficient, each of the literals and its constituent powers are the factors. This means that $6x^2 y^3$ has $1, 2, 3, x, x^2, y, y^2$ and y^3 as factors.

Concept 79

For a simple expression we use the following technique:
1. We find the common factors across the terms of the expression.
2. We divide each of the terms with these factors.
3. We enclose the quotients within brackets and multiply this expression with the common factors.

For example, in the expression, $ax^2 + bx$, x is a common factor between ax^2 and bx. Therefore we can write $ax^2 + bx = x(ax + b)$. Therefore the factors of $ax^2 + bx$ is x and $(ax + b)$

We can also arrange the expression into factors if the terms can be grouped to identify the common factors. Let us consider an example to understand this well.

$$x^2 + xy + xz + yz = x(x + y) + z(x + y) = (x + y)(x + z)$$

x is the common factor between the first two terms and z is the common factor between the last two terms. On resolving the expression into factors, another common compound factor emerges between the $x(x + y)$ and $z(x + y)$, which is $(x + y)$. Therefore, we can repeat the process of resolution one more time. Let us put all the steps together to ensure that we get the flow of the thought.

$$x^2 + xy + xz + yz = x(x + y) + z(x + y) = (x + y)(x + z)$$

Concept 80

The resolution of factors in binomial expressions is the next concept of interest. Let us start the discussion with a simple case of two compound expressions creating a trinomial expression.

Multiplying $(x + a)(x + b)$, we get $x^2 + x(a + b) + ab$.

Therefore, $(x + a)(x + b) = x^2 + x(a + b) + ab$

Or the factors of $x^2 + x(a + b) + ab$ are $(x + a)(x + b)$.

For example, the process of resolving $x^2 + 8x + 15$ is as follows.

$$x^2 + 8x + 15 = (x + a)(x + b) = x^2 + x(a + b) + ab$$

Therefore $ab = 15$ and $(a + b) = 8$. By inspection, we can see that $a = 5$ and $b = 3$

Therefore $x^2 + 8x + 15 = (x + 5)(x + 3)$

Concept 81

The difference of two squares can be resolved into factors by using the following formula:

$$a^2 - b^2 = (a + b)(a - b)$$

The product of sum and difference of two terms is equal to the difference of their squares. Conversely, the difference of two squares is equal to the product of the sum and difference of two quantities. The sum and difference constitutes the two factors of interest.

Concept 82

Two more formulas of interest are:

$$a^3 + b^3 = (a+b)(a^2 - ab + b^2)$$
$$a^3 - b^3 = (a-b)(a^2 + ab + b^2)$$

The left hand sides of the equations represent the sum and difference of cubes. The right hand side tells us how to resolve them into factors.

Concept 83

Using the results from the concepts we just saw, the ones dealing with factorization of sum and difference of squares and cubes respectively, we can factorize differences of larger powers. One such example is shown below.

$$a^6 - b^6 = (a^3 + b^3)(a^3 - b^3) =$$
$$(a+b)(a^2 - ab + b^2)(a-b)(a^2 + ab + b^2)$$

This is a good example to highlight how one can systematically approach problems in factorization.

Concept 84

Let us now consider the generic case of a trinomial.

$$ax^2 + bx + c = (px + q)(rx + s) = prx^2 + x(ps + qr) + qs$$

Therefore, we have $pr = a$; $ps + qs = b$; and $qs = c$. We can use these relationships to make intelligent conclusions about the factors and their coefficients p, q, r and s.

To start the process, check if a, b or c are prime numbers.

If a is a prime number, then either p or r or both are equal to 1. And the other variable is equal to a. We plug these numbers to determine the other coefficients and complete the solution.

Usually, a couple of intelligent guesses should do the trick ☺

Concept 85

With a good grasp of techniques for factorization, we can look at determining the HCF of a set of algebraic expressions. HCF is defined as the expression of the highest dimension that divides a set of algebraic expressions.

One way to determine the HCF of algebraic expressions is to factorize all the given expressions. The common factor of the highest dimension across all the expressions is the HCF. We can use all the techniques of factorization that we have seen so far.

Concept 86

If the usage of simple techniques of factorization is not obvious, we can use the technique similar to what we do in the case of determining HCF of a set of numbers.

This is based on two basic observations.
1. If an expression contains a certain factor, any multiple of the expression is also divisible by that factor.
2. If two expressions contain a common factor, it will divide the sum and difference of the expressions; and also multiples of the sums and difference of those expressions.
3. We eliminate the simple factors prior to applying these rules. This makes things simpler.

This is appears to be complicated when we write these statements in English. Let us demonstrate the concept using an example for improved clarity.

Find the HCF of $24x^4 - 2x^3 - 60x^2 - 32x$ and $18x^4 - 6x^3 - 39x^2 - 18x$

Step 1: Eliminate simple factors from the expressions.
1. $24x^4 - 2x^3 - 60x^2 - 32x = 2x(12x^3 - x^2 - 30x - 16)$
2. $18x^4 - 6x^3 - 39x^2 - 18x = 3x(6x^3 - 2x^2 - 13x - 6)$.
3. The HCF of $(12x^3 - x^2 - 30x - 16)$ and $(6x^3 - 2x^2 - 13x - 6)$ will give us the required answer.

Common Factors	Expression – 1	Expression - 2	Common Factors

$2x$
Expression
-2

$12x^3 - x^2 - 30x - \quad 6x^3 - 2x^2 - 13x \quad 2x \times (3x^2 - 4x -$

$12x^3 - 4x^2 - 26x - \quad 6x^3 8x^2 - 8x$

$x \times (3x + 2) \quad 3x^2 - 4x - 4 \qquad 6x^2 - 5x - 6 \qquad 2x(3x^2 - 4x - 4$
$)$
$\qquad\qquad\qquad\qquad\qquad\qquad\qquad\qquad$ 2 x (3x2 – 4x – 4)

$\qquad\qquad 3x^2 + 2x \qquad\qquad 6x^2 - 8x - 8$

$-2x(3x + 2 \quad -6x - 4 \qquad\qquad 3x + 2$

$\qquad\qquad -6x - 4$

$\qquad\qquad 0$

Therefore $3x + 2$ is the HCF.

Concept 87

The combined techniques of factorization and HCF can be used for reducing fractions to its lowest terms.

1. We factorize the numerator and denominator. This will throw up common factors if any. Such common factors can be cancelled out, this ensuring that our fraction is in its lowest terms.
2. If common factorization techniques are not obvious; we determine the HCF of numerator and denominator.
3. We divide the numerator and denominator by the HCF to determine the other factor. Thus we can reduce the fraction to its lowest terms.
4. Factorization can also be applied to reduce fractions to their lowest terms during the process of multiplication and division of fractions.

Concept 88

The process of determining the least common multiple of fractions is equally straightforward. We factorize the expressions and determine the multiple of common factors that divides the expression without leaving a reminder.

Concept 89

If HCF and LCM are two expressions for highest common factor and least common multiple of two expressions, then
1. Product of the two expressions = LCM x HCF
2. Therefore, LCM = Product of two expressions / HCF
3. And, HCF = Product of two expressions / LCM

Concept 90

An equation of the second degree is also known as the quadratic equation. The general form of a quadratic equation is $ax^2 + bx + c = 0$.

Here is a mathematical trivia. A quadratic equation of the form $ax^2 + bx + c = 0$ is also known as adfected quadratic; while a quadratic equation of the form $ax^2 + c = 0$ is also known as pure quadratic. We will drop the reference to the term "adfected" and simple refer to these equations as general quadratics.

We will commence our discussion of quadratic equations with pure quadratics.

Concept 91

A pure quadratic equation is of the form $ax^2 + c = 0$. We are solving for the value of x given x^2. In other words, we are dealing with finding the roots of an equation. Therefore, it is important for us to quickly recall the rule of signs for determining roots.

An even root of any positive term or quantity can be either positive or negative. This is because product of two positive terms or two negative terms is always positive. Therefore, the root of a quadratic [an even root] can be positive or negative.

If $x^2 = 25$; then $x = +5$ or $x = -5$ will satisfy the equation. This is represented by $x = \pm 5$ and is read as "x equals plus or minus five".

The roots of a pure quadratic of the type $ax^2 + c = 0$ are

$$x = \pm \sqrt{\left(\frac{-c}{a} \right)}$$

This formula is usually not memorized. We simply transpose the numbers and constant terms to the right hand side of the equation and collect all the terms with second power of the unknown to the left hand side of the equation. The solution to the determining the root of the unknown then becomes obvious and straightforward.

For example,

$$\frac{9}{(x^2-27)} = \frac{25}{(x^2-11)}$$

$$9(x^2-11) = 25(x^2-27)$$

$$9x^2 - 99 = 25x^2 - 675$$

$$16x^2 = 572$$

$$x^2 = 36$$

Therefore, $x = \pm 6$

The process of arriving at the solution to a pure quadratic from first principles is simple and straightforward.

Concept 92

Although a few expressions are not strictly pure quadratics, we can still use the above technique in handling general quadratic equations.

Consider $(x+a)^2 = b$

Then we can conclude that $x + a = \pm\sqrt{b}$

Therefore the solution to the equation is $x = -a \pm \sqrt{b}$

In other words, if we can convert a general quadratic into an equation of the form $(x+a)^2$ by completing the square, we can solve the roots like we did before.

Concept 93

Let us look at a general quadratic equation.

$$ax^2 + bx + c = 0$$

$$x^2 + \left(\frac{b}{a}\right)x + \left(\frac{c}{a}\right) = 0 \cdots \text{dividing both sides of the equation by}$$

coefficient of x^2

$$x^2 + \left(\frac{b}{a}\right)x = -\left(\frac{c}{a}\right) \cdots \text{transposing}\left(\frac{c}{a}\right) \text{ to the right hand side of the}$$

equation

Let us complete the square on the left hand side of the equation.

$$x^2 + 2.\left(\frac{b}{2}\right)x + \left(\frac{b}{2a}\right)2 = -\left(\frac{c}{a}\right) + \left(\frac{b}{2a}\right)2 \cdots \text{adding} + \left(\frac{b}{2a}\right)2 \text{ to both sides}$$

$$\left(x + \left(\frac{b}{2a}\right)\right)^2 = \left(\left(\frac{b}{2a}\right)^2 - \left(\frac{c}{a}\right)\right)$$

$$= \frac{(b^2 - 4ac)}{4a^2}$$

$$x + \left(\frac{b}{2a}\right) = \pm\sqrt{\frac{(b^2 - 4a)}{(2a)}}$$

$$x = \frac{[-b \pm \sqrt{(b^2 - 4a)}]}{(2a)}$$

These are the two roots of a general quadratic equation.

Concept 94

Ratio is simply a relationship between two similar quantities. The nature of comparison is about what part of one is the other; or what multiple of one is the other. This is written as $a : b$. The first part a is called antecedent and the other part b is called consequent.

The ratio is said to be of greater inequality, equality or lesser inequality when antecedent is greater than, equal to or less than the consequent respectively.

Concept 95

A ratio a:b is the same as the fraction a/b.

Concept 96

The laws of fraction are applicable to ratios as well.

$$\frac{a}{b} = \frac{(ma)}{(mb)} \text{ for all } m \neq 0$$

Therefore $a = b = ma + mb$ for all $m \neq 0$

In other words, the value of a ratio remains unaltered if the antecedent and consequent of a ratio are divided by the same quantity.

Concept 97

Let us consider two ratios $a : b$ and $x : y$

$a : b = \dfrac{a}{b} = \dfrac{ay}{by}\cdots$ value is not altered if we multiply and divide by the same quantity

$x : y = \dfrac{x}{y} = \dfrac{bx}{by}\cdots$ value is not altered if we multiply and divide by the same quantity

Therefore
1. $a : b > x : y$ when $ay > bx$
2. $a : b = x : y$ when $ay = bx$
3. $a = b < x : y$ when $ay < bx$

The above relationship is also known as rule of cross products.

Concept 98

If $a : b, p : q$ and $x : y$ are given ratios, then $apx : bqy$ is called the compounded ratio. We simply take the product of antecedents and consequents to determine the compounded ratio of two or more ratios.

Concept 99

When $x:y$ is compounded by itself, we get $x^2:y^2$. This is called the duplicate ratio of $x:y$. And, $x^3:y^3$ is called triplicate of the ratio $x:y$. $\sqrt{x}:\sqrt{y}$ is called sub-duplicate ratio of $x:y$

Concept 100

If $a+b=c+d=e:f$, then

$ax+by+cz:bx+dy+ez = a:b=c:d=e:f$.

In other words, when a series of fractions or ratios are equal to one another, then each of the fractions is equal to the sum of all numerators divided by sum of all denominators.

Concept 101

When two ratios are equal, then the four quantities are said to be in proportion.

If $a:b=c+d$, then a,b,c and d are in proportion. This is written as $a:b::c:d$, and read out as a is to be is as is to c is to d. a and d are called extremes and b and c are known as means.

It may be noted that when four quantities are in proportion, then product of means is equal to product of extremes.

Concept 102

Three quantities a,b and c are said to in continued proportion if

$a:b::b:c$ or $\dfrac{a}{b}=\dfrac{b}{c}$

The product of means = Product of extremes

$b^2 = ac$

b is called the mean proportion of a and c.

Concept 103

If three quantities are in continued proportion, then the ratio of the first and third quantity is the duplicate ratio of first and second quantity. If $a : b : c$, then $a : c :: a^2 : b^2$

Concept 104

If $a : b :: c : d$; and $e : f :: g : h$, then $ae : bf :: cg : dh$

As a consequence, if $a : b :: c : d$; and $b : x :: d : y$, then $a : x :: c : y$

Concept 105

1. If $a : b :: c : d$ then $b : a :: d : c$
2. If $a : b :: c : d$ then $a : c :: b : d$
3. If $a : b :: c : d$ then $(a + b) : b :: (c + d) : d$
4. If $a : b :: c : d$ then $(a - b) : b :: (c - d) : d$
5. If $a : b :: c : d$ then $(a + b) : (a - b) :: (c + d) : (c - d)$

Concept 106

If $a : b :: c : d$, then

1. $ma : mb :: nc : nd$
2. $ma : nb :: mc : nd$
3. $an : bn :: cn : dn$
4. $pan : qbn = pcn : qdn$

Concepts in Euclidean Geometry

Concept 27: Opening Movies

1. Human thought process is complex. We gather information from our surrounding. We use this information to make "informed" decisions.

2. The process of using information to make decision is called "reasoning".

3. Mathematical reasoning is based on Mathematical building blocks, foundations, principles and observations. For example,

   a. If equals are added to equals, then the sums are equal.

   b. If equals are taken away from equals, then the remainders are equal.

   c. Halves of equals are equal to one another.

4. We will commence our journey into the World of Geometry.

Concept 28: Truths

1. **Axioms** are true by definition.

2. **Postulates** are statements that are true and which require no further proof.

3. **Theorem** is a statement whose truth can be established based on the axioms and other derived theorems.

4. **Corollary** is a statement, whose truth is established based on existing theorems. It is a proposition that can be derived, inferred or deduced from a theorem or an axiom, not requiring any further proof.

5. When a theorem is proved, we append the letters QED. QED stands for **quod era at demonstrandum**, or "which was to be proved".

6. Two geometric figures or quantities are similar, if their shapes are same, but the dimensions are different. A square of side 5 units is similar to another square of side 8 units.

7. Two geometric figures or quantities are congruent if their shapes and dimensions are identical. **Congruent** shapes fit exactly on top of one another after some movement.

8. This completes the overview of introductory concepts in Geometry. We will now look into each of the concepts systematically in the subsequent concepts.

Concept 29: Point

1. The concepts of a point, line and surface are familiar to all of us. We use this to describe objects or shapes that we see. We need to revisit these concepts and arrive at mathematical definitions for these shapes.
2. The **point** has a position, but no other feature, such as length, breadth, width height or thickness.
3. We can say that a **point** has a **position**, but **no magnitude**.
4. In order to represent a point, we take a sharp pencil and make a dot. We see that such a dot has features like thickness, length or breadth depending upon the sharpness of the pencil lead.
5. The smaller the dot, the closer we get to the concept of a point.

Concept 30: Line

1. A **line** has **length** associated with it, but **no breadth**.
2. Let us extend the idea we used while discussing the concept of point.
3. The figure traced by a moving point is a line. The sharper the pencil we use to trace the line, the closer we get to the concept of a line.
4. A line can be straight or curved.

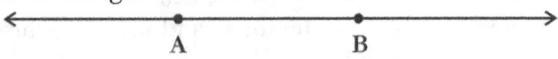

Fig 1: $\overleftrightarrow{AB}, \overrightarrow{AB}, \overrightarrow{BA}$ are rays. AB is a straight line.

5. A straight line is a line where there is no change in direction between any two points on it.
6. A curved line is characterized by change in direction between the points.

Fig 2: AB is a curved line

7. **AXIOM: There can be only one straight line joining two points.**

8. As a consequence, **two straight lines cannot enclose a space.**

9. To bisect a line is to divide a line into two equal parts. And to trisect is to divide the line into three equal parts.

10. If a point O moves from point A to point B on a straight line AB, then it passes through one point which bisects the line AB.

11. Every finite line segment has a point of bisection.

Concept 31: Rays

1. Lines that extend on either or both directions of a straight line are called **rays**.

Fig 3: $\overrightarrow{AB}, \overrightarrow{AB}, \overrightarrow{BA}$ are rays. AB is a straight line.

Concept 32: Surface

1. This brings us to the concept of a surface. A surface has a length and breadth, but no thickness.

2. A **surface** is bounded by lines.

3. A line is bounded at points; therefore lines meet at point. These points are called points of intersection.

4. A **plane** is a flat surface. A straight line drawn between two points on a plane, lies on the plane completely.

5. A plane consists of atleast 3 points that are all not on the same straight line.

6. Any three points that are not located on the same straight line are contained in a plane.

7. A straight line and a point not on it is contained by one and only one plane.

8. Three collinear points are contained in more than one planes. In fact, in an infinite number of planes.

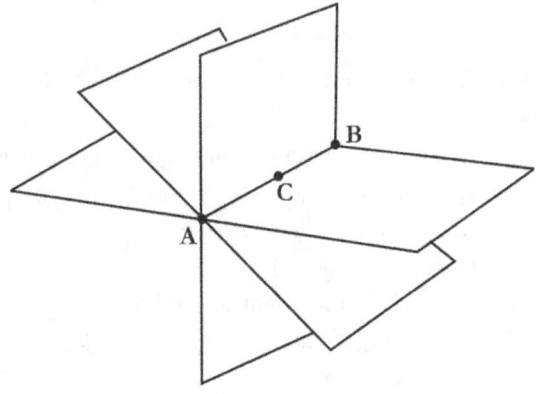

Fig 4: Three collinear points are contained in infinite number of planes

9. When two points are on a plane, then the line containing the two points is also on the same plane.

10. When two planes intersect, then their intersection is a straight line.

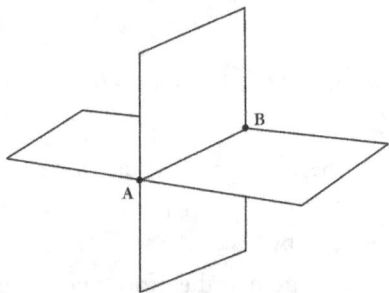

Fig 5: When two planes intersect, the intersection is a straight line

Concept 33: Solid

1. A solid has a notion of length, breadth and thickness.

2. A solid is bounded by surfaces.

3. The notion of space is generic, yet critical. Space is a set of all points and it extends indefinitely in all directions and dimensions.

4. Space contains atleast four points that are not on the same plane.

Concept 34: Angle

1. When two lines meet at a point, they form an **angle**.
2. The two lines are called arms of the angle.
3. The point of intersection is called the **vertex**.

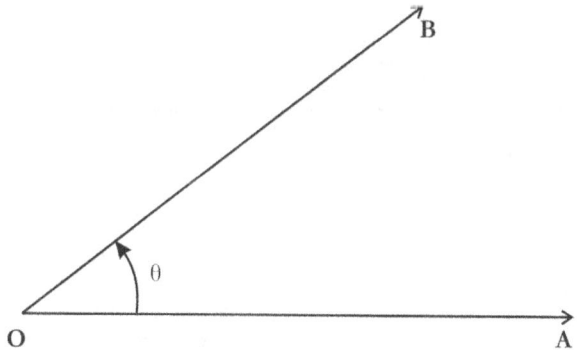

Fig 6:θ : angle; O: vertex; $\overrightarrow{OA}, \overrightarrow{OB}$: Arms of the angle

4. Measuring an angle:
    a. Consider two arms (straight lines), OA and OB, meet at point O.
    b. Fix the position of OA. Make OB lie on top of OA.
    c. Now, move OB from its original position to a new position.
    d. The size of AOB is measured by the amount of ``turning'' of the revolving arm OB.
    e. Clearly, the "amount of turn" is not dependent on the length of OA and OB. Therefore, the angle is independent of length of the line.
5. The angles on either of a common arm is called adjacent angles. Consider three arms, OA, OB, OC meeting at point A. $\angle AOB$ and $\angle BOC$ are called adjacent angles.
6. Clearly, $\angle COA = \angle COB + \angle BOA$
7. When two straight lines AB and CD intersect at point O; the angles $\angle AOC$ and $\angle BOD$ are vertically opposite angles.

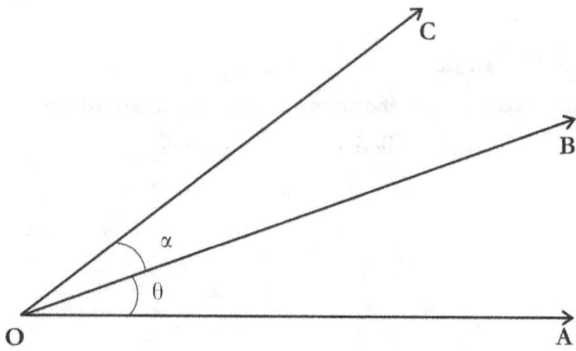

Fig 7: ∠AOB, ∠BOC are adjacent angles

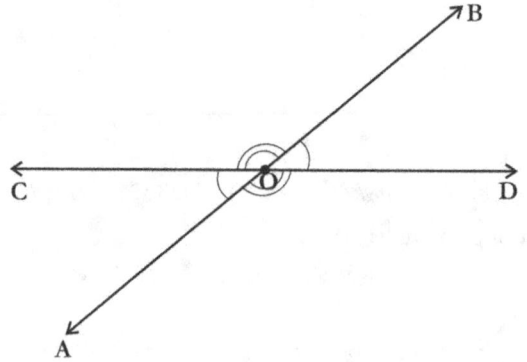

Fig 8: ∠AOD, ∠BOC : Vertically opposite angles ∠AOC, ∠BOD : Vertically opposite angles; ∠AOD = ∠BOC and ∠AOC = ∠BOD

8. When a line OC stands on a straight line AB, such that the adjacent angles ∠COA = ∠COB. The line OC is said to be **perpendicular** to AB. This angle is also called a right angle.

9. **AXIOM:** If O is a point on a straight line AB, and a line OC turns about point O, from OA to OB, then there is one and only one point when OC is perpendicular to AB.

10. All right angles are equal.

11. We divide the right angle into 90 equal parts. Each of these parts is called a degree. A right angle is equal to 90 degree. It is represented as 90^0.

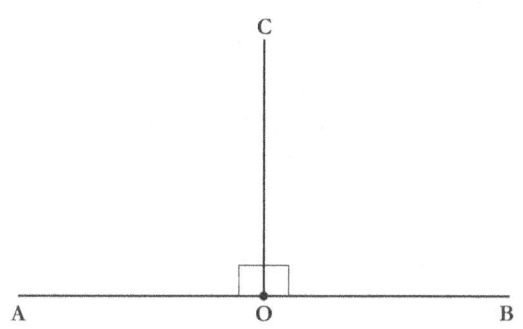

Fig 9: *AB is a straight line;* $\angle COA = \angle COB \Rightarrow OC \perp AB$;
$$\angle COA = \angle COB = \frac{\pi}{2} ; \angle AOB = \pi$$

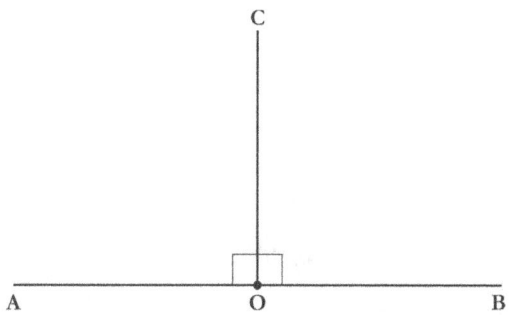

Fig 10: *AB is a straight line;* $\angle COA = \angle COB \Rightarrow OC \perp AB$;
$$\angle COA = \angle COB = \frac{\pi}{2} ; \angle AOB = \pi$$

12. The $\angle AOB$ is equal to two right angles. Therefore, angle of a straight line is two right angles. The straight angle is 180^0.

13. An angle less than a right angle is called acute angle. An acute angle is less than 90^0.

14. An angle greater than a right angle and less than two right angles is called obtuse angle. An obtuse angle is between 90^0 and 180^0.

15. An angle greater than two right angles is called reflex angle. A reflex angle is between 180^0 and 360^0.

16. Consider an angle $\angle AOB$. If a line OC turns about point O, from OA to OB, it passes through one position where it bisects $\angle AOB$

17. This means, every angle, has a line of bisection.

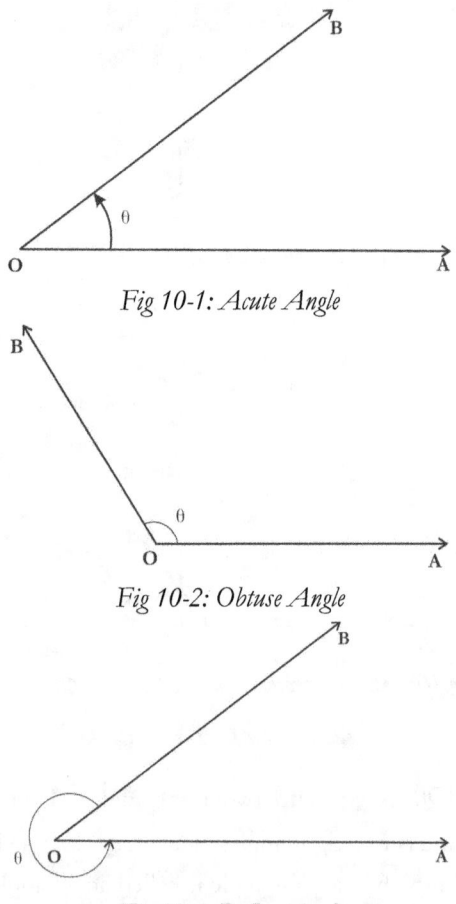

Fig 10-1: Acute Angle

Fig 10-2: Obtuse Angle

Fig 10-3: Reflex Angle

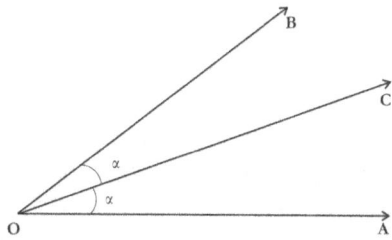

Fig 11: If $\angle AOC = \angle COB = \alpha$; $O\vec{C}$ *is the line of bisection*

18. Every point on the angular bisector is equidistant from the sides of the angle. The distance measured is the perpendicular distance. [*Fig:12*]

19. If a point is equidistant from the arms of an angle, then the point lies on the angular bisector.

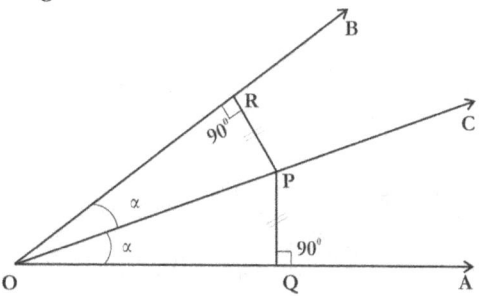

Fig 12: If a point is equidistant from the arms of an angle, then the point lies on the angular bisector.

20. Every point on the perpendicular bisector is equidistant from the end point of the line.

21. An angle is said to be trisected if the rays / lines contains the vertex and split the angle into three equal adjacent angles.

22. Complementary angles are angles whose sum equals one right angle.

23. If two angles are complementary to the same angle, then the two angles are equal.

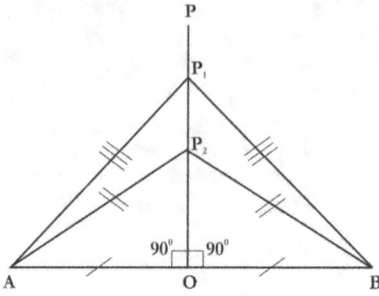

Fig 13: Every point on the perpendicular bisector is equidistant from the end point of the line.

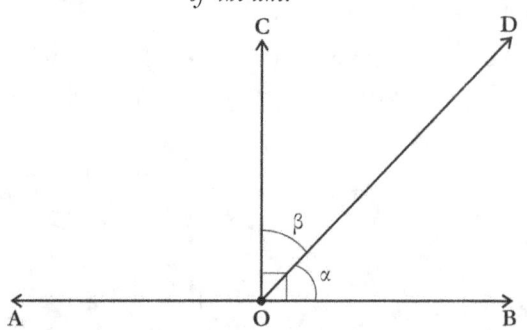

Fig 14: $\angle COB = \dfrac{\pi}{2} = \alpha + \beta$; $\angle BOD, \angle DOC$ *are complementary.*

24. If two angles are complementary to congruent angles, then they are congruent.

25. Supplementary angles are angles whose sums are equal to two right angles.

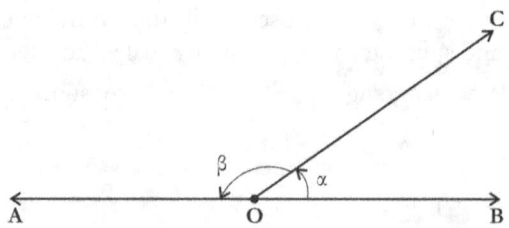

Fig 15: $\angle AOB = \pi = \alpha + \beta$; $\angle BOC, \angle COA$ *are supplementary*

26. If two angles are supplementary to the same angle, then the two angles are equal.

27. If two angles are supplementary to congruent angles, then they are congruent.

28. When two lines intersect, the point of intersection is called the vertex. The angle subtended at the vertex is called vertical angles.

29. Vertical angles are congruent.

30. Adjacent vertical angles are supplementary.

Concept 35: Circle

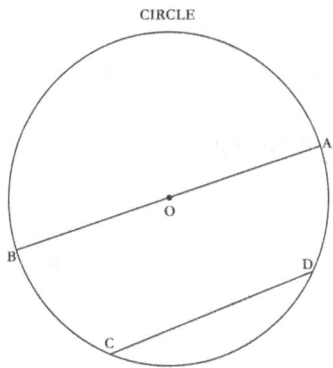

Fig 16: O: center; OA: radius, BOA: diameter; CD: chord

1. Any portion of a plane surface that is bounded by one or more lines is called a plane figure.

2. A circle is a plane figure that is traced by a point that is at a constant distance from a fixed reference point.

3. The fixed reference point is called center of the circle.

4. The constant distance between the point and the center of the circle is called radius of the circle.

5. The straight line distance from the center of the circle to any point on its boundary is therefore the same as the radius of the circle.

6. The bounding line is called the circumference of the circle.

7. The diameter of a circle is simply a straight line drawn through the center of the circle terminated both sides at the circumference of the circle.

8. The diameter of the circle is twice the radius of the circle.

9. The line that ends on the circumference of a circle, but which does not pass through the center of the circle is called a chord.

10. Any portion of the circumference is called an arc of the circle.

11. A semicircle is simply an arc that is bounded by a diameter of the circle.

Concept 36: Properties of Lines and Angles

We will now dive deep into the concepts of lines and angles.

1. Points on the same line are collinear. Collinear points are on the same straight line.

2. Noncollinear points are not on the same straight line.

3. Any three noncollinear points are coplanar. They are always on some plane.

4. Intersecting lines share one and only one point in common.

5. Intersecting lie on one plane.

6. Perpendicular lines are lines that intersect and form a right angle at the point of intersection.

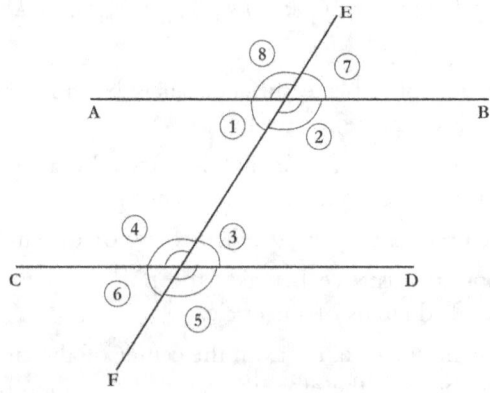

Fig 17: AB is parallel to CD; EF is transversal. Angles 1,2,3,4 are internal angles; Angles 5,6,7,8 are external angles

7. Only one perpendicular can be drawn to a line, from a point that is not on the straight line.

8. The shortest distance from any point to a line or a plane is perpendicular distance.

9. A perpendicular bisector of a line segment is simply a perpendicular at its midpoint.

10. A point on the perpendicular bisector is equidistant from the end points of the line segment.

11. Conversely, a point equidistant from the end points of a line segment lies on the perpendicular bisector of the line segment.

12. A transversal is a line that intersects two or more coplanar lines at different points, one each on each of the lines.

13. Parallel lines are coplanar.

14. Parallel lines do not intersect each other. In other words, parallel lines meets at infinity.

15. The distance between two parallel lines remains the same everywhere.

16. Through a point which is not the line, exactly one parallel line can be drawn.

17. If three or more parallel lines cut off equal segments on a transversal, then they will cut off equal segments on every other transversal that they share. This is simply restating the property of a set of parallel lines.

18. Interior angles are formed between two lines and a transversal, such that the regions of the angles are between the two lines.

19. Alternate interor angles have different vertices and are on opposite side of the transversal.

20. If the lines are parallel, then the alternate interior angles are equal.

21. If the alternate interior angles are equal, then the two lines are parallel.

22. As opposed to an alternate interior angles, the same-side interior angles have different vertices but on the same side of the transversal.

23. If the lines are parallel, then the same-side interior angles are supplementary.

24. If the same-side interior angles are supplementary, then the lines are parallel.

25. Exterior angles are formed between two lines and a transversal, such that the regions of the angles are outside of the region between the two lines.

26. Alternate exterior angles are exterior angles with different vertices and on opposite sides of the transversal.

27. If the lines are parallel, then the alternate exterior angles are equal.

28. If the alternate exterior angles are equal, then the two lines are parallel.

29. As opposed to an alternate exterior angles, the same-side exterior angles have different vertices but on the same side of the transversal.

30. If the lines are parallel, then the same-side exterior angles are supplementary.

31. If the same-side exterior angles are supplementary, then the lines are parallel.

32. Corresponding angles are angles with different vertices, on the same side of the transversal, and are in the same relative position as well.

33. In a pair of corresponding angles, one of them is interior angle and the other is exterior.

34. If the lines are parallel, then the corresponsing angles are equal.

35. If the corresponding angles are equal, then the lines are parallel.

36. If the lines are parallel and all the interior and exterior angles are equal, then the transversal is perpendicular.

37. If a transversal is perpendicular to one of the parallel lines, then it is perpendiculr to the other line too.

38. Skewed lines are non-coplanar, travel in different directions and never intersect.

39. A line segment is simply the a line with two defined end points. It can also be defined as a set of all collinear points between two defined end points. Line AB is the same as line BA.

Concept 37: Polygons

1. Geometric shapes formed by line segments that intersect at end points and enclose one and only one closed region between them is called a Polygon.

2. A polygon of 3 sides is called a triangle.

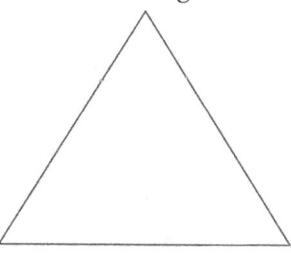

Fig 18: Triangle

3. A polygon of 4 sides are square, rectangle, parallelogram, rhombus, trapezium and quadrilateral. Quadrilateral is the most generic 4 sided polygon; all the others are quadrilaterals with specific properties.

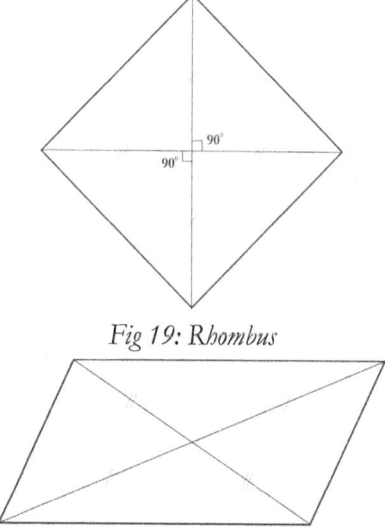

Fig 19: Rhombus

Fig 20: Parallelogram

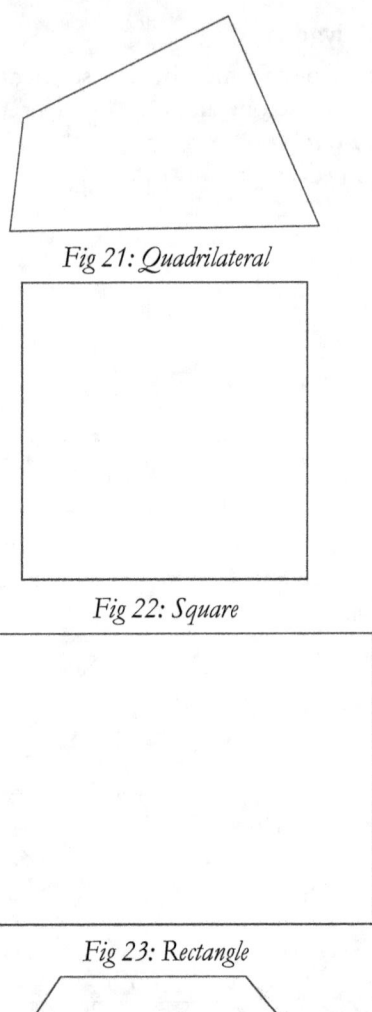

Fig 21: Quadrilateral

Fig 22: Square

Fig 23: Rectangle

Fig 24: Trapezium

4. Similarly, a 5-sided polygon is pentagon, 6 sided one is hexagon and so on.

5. Each of the intersecting end point of the sides of the polygons is called vertex.

6. The lines connecting vertices of the polygons, which are not the sides of the polygons are called diagonals.

7. There are $n(n-1)/2$ diagonals in a n-sided polygon.

8. The interior of a polygon is simply the region enclosed within the sides of the polygon.

9. The interior angles of a polygon are formed at each vertex of the polygon, interior to the polygon.

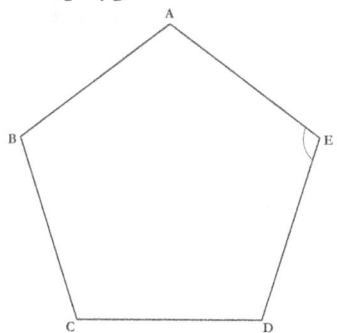

Fig 25: Polygon with A,B,C,D,E as vertices; ∠AED : internal angle

10. A concave polygon has atleast one interior angles which is greater than two right angles.

11. A convex polygon has no interior angles greater than two right angles.

12. The number of vertices, interior angles and sides of a polygon are all equal.

13. The sum of the interior angles of a n-sided polygon is $2(n-2)$ right angles. Pick any vertex of the polygon. You can draw all the diagonals from that vertex. You will find $n-2$ triangles. Since the sum of angles of a triangle is 2 right angles, we can conclude that the sum of interor angles of a polygon is $2(n-2)$ right angles.

14. Exterior angles are formed by extending the sides. An exterior angle is formed by one side of the polygon and an extended side, at the vertex.

15. The sum of exterior angles of a polygon using one exterior angle at each vertex is four right angles.

16. Regular polygons are polygons, whose sides are equal in length and the interior angles are all equal to one another.

17. In the subsequent concept we will focus our attention on simplest polygon; a polygon with 3-sides---a triangle.

Concept 38: Properties of Triangles

1. Triangles are polygons with three sides, three vertices and three interior angles. The symbol for a triangle is Δ .

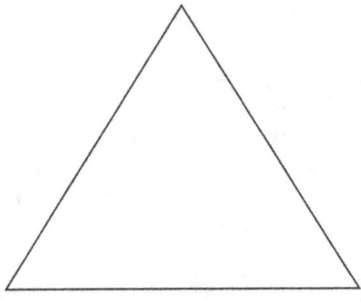

Fig 26: Triangle

2. The interior angles of a triangles are commonly referred to as the angles of the triangle.

3. The length of the perpendicular from a vertex to the opposite side is the height of the triangle. The opposite side is known as the base of the triangle.

4. The base of the triangle is not necessarily the side which the triangle lies on. It need not be the bottom of the triangle. The side opposite to a vertex is called the base of the triangle.

5. Every triangle has three vertices, three heights and three bases.

6. Every perpendicular that determines the height of the triangle has a different base associated with it.

7. There are 3 types of triangles.

    a. Scalene Triangle : All the three sides are of different lengths.

    b. Isosceles Triangle : Two of the sides are equal in length.

c. Equilateral Triangle : All three sides are equal in length.

d. A equilateral triangle is also an isosceles triangle.

e. Obtuse Angle Triangle : When one of the interior angles is an obtuse angle.

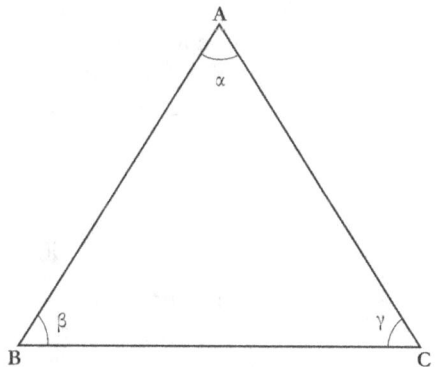

Fig 27: Isosceles triangle; AB=AC; ∠ABC = ∠ACB = α

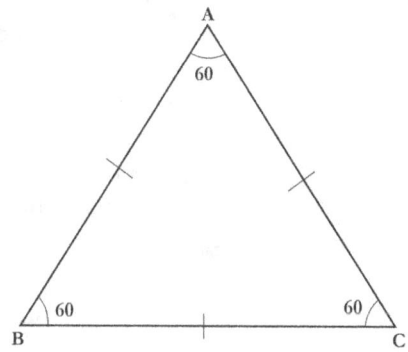

Fig 28: Equilateral triangle; AB=BC=CA;

$$\angle ABC = \angle BCA = \angle CAB = \frac{\pi}{3}$$

f. Right Angled Triangle : When one of the interior angles is equal to 90^0 or a right angle.

    i. The side opposite to the right angle is called hypotenuse. It is also the longest side in a right angled triangle.

    ii.   The other two sides are called legs of the triangle.

    iii. Pythogoras Theorem: The square on the hypotenuse of a right angled triangle is equal to the sum of the squares on the other two sides.

    iv. If the square on the longest side of a triangle is equal to the sum of the squares on the other two sides, then the triangle is a right angled triangle.

    v.   In a $45^0 : 45^0 : 90^0$ triangle, the sides are in the ratio of $1 : 1 : \sqrt{2}$.

    vi. In a $30^0 : 60^0 : 90^0$ triangle, the sides are in the ratio of $1 : \sqrt{3} : 2$.

    vii. The midpoint of the hypotenuse of a right angled triangle is equidistant from all the three vertices.

    viii. When a perpendicular is drawn to the hypotenuse of a right angled triangle, the two triangles are similar to each other and to the right angled triangle.

    ix. The perpendular is the geometric mean of the segments of the hypotenuse.

    x.   Each leg of the right angled triangle is the geometric mean of the hypotenuse and the line segment of the hypotenuse adjacent to it.

  g. Acute Angled Triangle : When all the three interior angles are acute angles.

  h. Equiangular Triangle : When all the three interior angles are equal. An equiangular triangle is an equilateral triangle.

8. The sum of the interior angles of a triangle is equal to two right angles or 180^0.

9. If two angles of a triangle are equal to the two angles of another triangle, then the third angle of both triangles are equal.

10. Each angle of an equilateral triangle is 60^0.

11. There can be no more than one right angle or obtuse angle in a triangle.

12. The acute angles of a right angle triangle are complementary. The sum of these angles are 90^0.

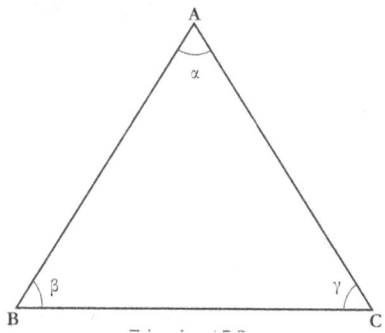

Fig 29: Sum of internal angles of a triangle is 180⁰; $\alpha + \beta + \gamma = \pi$

13. The exterior angle of a triangle is equal to the sum of the two interior angles, which do not share the same vertex.

14. The sum of any two sides of the triangle is greater than the third side.

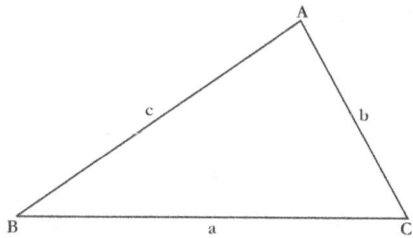

Fig 30: Sum of any two sides of a triangle is greater than the third side; Therefore, a+b>c; c+a>b; b+c>a

15. SSS Postulate: If three sides of one triangle are equal in length to three sides of another triangle then the triangles are congruent.

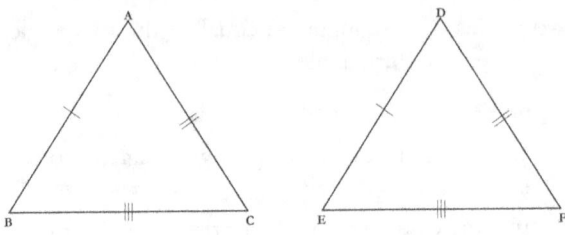

Fig 31: SSS Postulate

16. SAS Postulate: If two sides and one angle of a triangle are equal to two sides and one angle of another triangle, then the triangles are congruent.

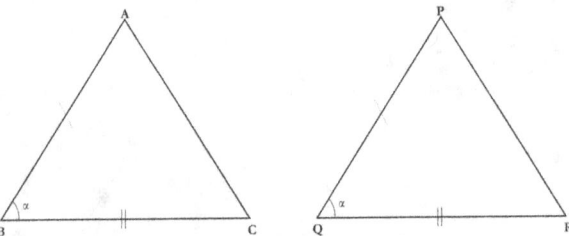

Fig 32: SAS Postulate

17. ASA Postulate : If two angles and one side of a triangle are equal to two angles and one side of another triangle, then the triangles are congruent.

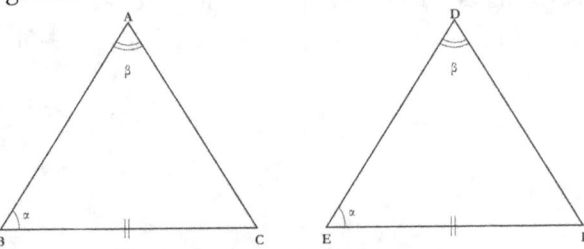

Fig 33: ASA Postulate

18. AA Rule : Two angles of one triangle are the same as two angles of another triangle, then the triangles are similar, and not necessarily congruent.

19. If two sides of a triangle are equal, then the interior angles opposite to those sides are equal.

20. If two angles of a triangle are equal, then the sides opposite to those angles are equal.

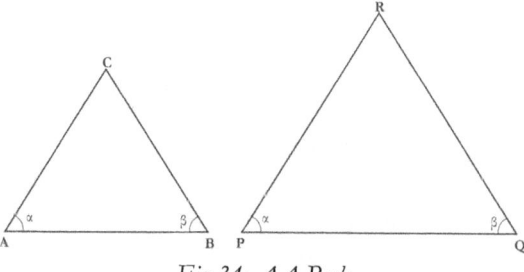

Fig 34: AA Rule

21. All angles of an equilateral triangle are equal to 60^0.

22. The angular bisector of the vertex of an isosceles triangle is the perpendicular bisector of the base.

23. If the hypotenuse and a side of a right angled triangle is equal to the hypotenuse and a side of another right angled triangle, then the two right angled triangles are congruent.

24. If two sides of a triangle is equal to two sides of another triangle; and the included angle of one triangle is greater than the other, then the opposite side of the included angles of one triangle is greater than the other.

25. If two sides of a triangle are equal to two sides of another triangle, and the third side is of greater than the other, then the included angle of one triangle is greater than the included angle of the other triangle.

Concept 39: Properties of Quadrilaterals

1. uadrilateral is a four sided polygon.

2. The sum of the internal angles of a quadrilateral is 4 right angles or 360^0.

3. If the diagonals of a quadrilateral bisect each other, then the quadrilateral is a parallelogram.

4. A square, rectangle and rhombus are parallelograms.

5. Rhombus is a parallelogram whose diagonals bisect perpendicularly.

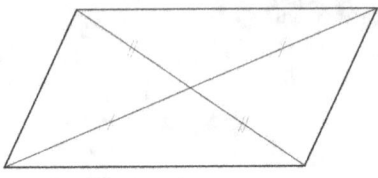

Fig 35: Parallelogram

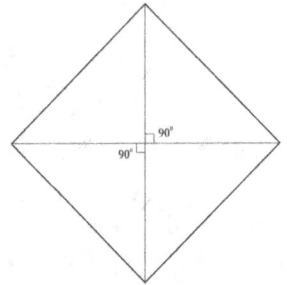

Fig 36: Rhombus

6. The opposite internal angles of a rhombus are equal.

7. Square is a rhombus. Rectangle is not necessarily a rhombus.

8. Square is a rectangle whose sides are all equal.

9. Diagonals of a rhombus are perpendicular.

10. Rhombus ∩ Rectangles = Squares

11. 4-side Polygons = Parallelograms ∪ Trapezoids ∪ Quadrilaterals

12. (Parallelograms ∪ Trapezoids) ⊂ Quadrilaterals

13. (Rectangles ∪ Squares ∪ Rhombus) ⊂ Parallelograms

Concept 40: Circles

1. A circle is a set of points which are equidistant from a fixed point called the **center** of the circle.

2. The distance between the center of the circle to the boundary of the circle is called **radius** of the circle.

3. The boundary or the edge of the circle is called the **circumference** of the circle.

4. The center of the circle lies inside the circle; it is not a point on the circumference of the circle.

5. A **chord** is a coplanar straight line segment that ends on the circumference on both sides. If the line extends on both sides indefinitely, it is called the **secant**. Secant therefore intersects the circumference at two points.

6. A **diameter** of the circle is the chord that passes through the center of the circle.

7. A diameter of the circle is therefore twice the length of the radius of the circle.

8. Two circles are said to be congruent or equal if their radii are equal. They are also called equal circles.

9. Concentric circles are circles that share the same point as their center.

10. A **tangent** is a line that is coplanar with the circle and touches the circle at one point, called the **point of tangency**.

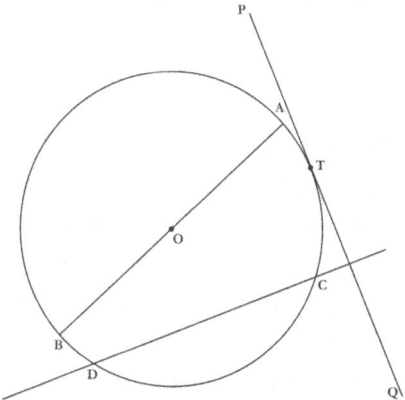

Fig 37: AOB: diameter; OA: radius; PTQ: tangent; T: point of tangency; CD=chord; $\overleftrightarrow{CD}$ =secant

11. If a line is tangent to a circle, then the radius whose end point is the same as the point of tangency is perpendicular to the tangent.

12. If a coplanar line is perpendicular to a radius at its outermost point, then the line is a tangent to the circle.

13. If two tangents intersect, then the line segments with the point of intersection and point of tangency as end points are equal in length.

14. The line segments drawn from a coplanar point outside of the circle to the points of tangency are equal in length.

15. A straight line that is tangent to two coplanar circles is called the **common tangent** of the two circles.

    a. Common Internal Tangents intersect between the two circles.

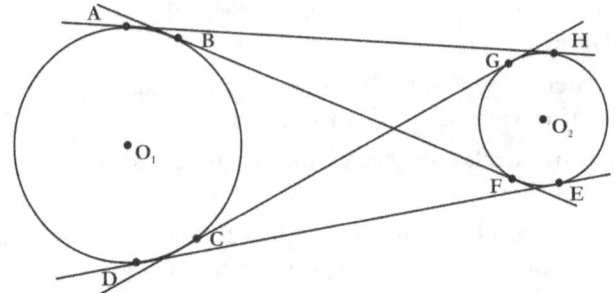

Fig 38: CG, BF: Internal tangents; AH, DE: External tangents

    b. Common External Tangents do not intersect betweent two circles.

    c. If common external tangents are parallel to each other, the circles have the same radius.

16. An arc is a part of the circumference of the circle.

    a. A semicircle is an arc that ends at the end points of a diameter.

    b. A minor arc is an arc that is lesser than a semicircle in length.

    c. A major arc is an arc that is greater than a semicircle in length.

    d. The end points of an arc subtend an angle at the center of the circle. This is called the **central angle**.

    e. The central angle of a semicircle is 180^0.

17. An **inscribed angle** is simply the angle whose vertex is on the circumference of the circle and the sides are the chords of the circle.

18. An **inscribed polygon** is formed when vertices of the polygon are on the circumference of the circle.

19. A **circumscribed polygon** is formed when the sides of the polygon are tangential to the circle.

Concept 41: Perimeter

1. Square of side $s = 4 \times s$

2. Rectangle of length l and breadth $b = 2 \times (l + b)$

3. Triangle of sides a, b and $c = a + b + c$

4. Circle of radius $r = 2 \times \pi \times r$

Concept 42: Areas

1. Square of side $s = s^2$

2. Rectangle of length l and breadth $b = lb$

3. Triangle of base b and height $h = \dfrac{1}{2} bh$

4. Trapezium of sides l_1 and l_2, and height $h = \dfrac{1}{2}(l_1 + l_2)h$

5. Parallelogram of side s, distance h between them $= sh$

6. Circle of radius $r = \pi r^2$

Concept 43: Surface Areas

1. Cube of side $s = 6s^2$

2. Cuboid of edges l, b and $h = 2(lb + bh + hl)$

3. Cyclinder of height h, radius $r = 2\pi r^2 + 2\pi rh$

4. Cone of radius r and slant height $s = \pi r^2 + \pi rs$

5. Sphere of radius $r = 4\pi r^2$

Concept 44: Volumes

Volume of :

a) Cube of edge $s = s^3$

b) Cuboid of edges l, b and $h = l \times b \times h$

c) Cyclinder of height h and radius $r = \pi r^2 h$

d) Cone of of height h and radius $r = \dfrac{1}{3}\pi r^2 h$

e) Rect. prism of length l, width w, height $h = \dfrac{1}{3}lwh$

f) Triangular prism=Area of the triangle $\times$ height

g) Sphere of radius $r = \dfrac{4}{3}\pi r^3$

Concepts in Plane Trigonometry

Concept 45: The Pythogorean Theorem
1. A triangle is a polygon consisting of three sides.
2. We use $\triangle ABC$ to represent a triangle with vertices A,B and C - having internal angles $\angle A$, $\angle B$ and $\angle C$.
3. The sides opposite to these angles will be referred using the corresponding lowercase letter. Therefore, a is the side opposite to $\angle A$, b is the side opposite to $\angle B$ and c is the side opposite to $\angle C$.
4. We will always assume that c is the longest side of the triangle; therefore $\angle C = 90°$ and C is the hypotenuse when $\triangle ABC$ is a right triangle.

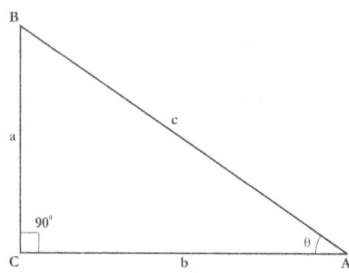

1. **Pythogorean Theorem:** In a right triangle, the square on the hypotenuse is equal to the sum of the squares on the other two sides.
2. This can be represented as follows: In a right triangle $\triangle ABC$, $c^2 = a^2 + b^2$, where c is the hypotenuse.

Concept 46: Units and Conversion
1. Sexagesimal Measures
   a. 60 Seconds = 1 Minute
   b. 60 Minutes = 1 Degree
   c. 90 Degrees = 1 Right Angle
2. Centesimal Measures
   a. 100 Seconds = 1 Minute
   b. 100 Minutes = 1 Grade
   c. 100 Grades = 1 Right Angle
3. Sexagesimal to Centesimal Conversion techniques
   a. 90 Sexagesimal degrees = 100 Centesimal degrees

b. $1° = \dfrac{10^g}{9}$

c. $1^g = \dfrac{9}{10}°$

4. Circular Measures

   a. Radian is the angle subtended at the center of the circle by an arc length equal to the radius.

   b. Independent of the radius, the length of the circumference of a circle always bears a constant ratio with its diameter. This constant ratio is π.

   c. In other words, the circumference is π times diameter or $2 \times \pi$ times radius. $2\pi \text{ radians} = 360° = 400^g$

5. If x is an angle in degrees, t is an angle in radians, then

   a. $\dfrac{\pi}{180} = \dfrac{t}{x} \Rightarrow t = \dfrac{\pi x}{180}$

   b. $x = \dfrac{180 t}{\pi}$

Concept 47: Definitions from a Right Triangle

Consider the right angled triangle ABC, shown in the figure. We can define the trigonometric ratios, and use this figure as a way of internalizing the definitions.

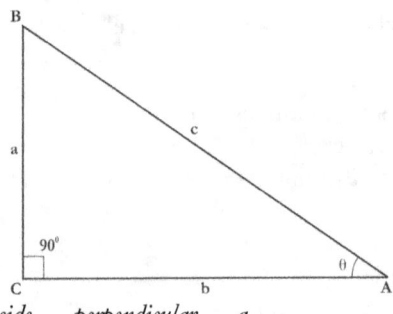

1. $\sin\theta = \dfrac{opposite\ side}{hypotenuse} = \dfrac{perpendicular}{hypotenuse} = \dfrac{a}{c}$

2. $\csc\theta = \dfrac{hypotenuse}{opposite\ side} = \dfrac{hypotenuse}{perpendicular} = \dfrac{c}{a}$

3. $\cos\theta = \dfrac{adjacent\ side}{hypotenuse} = \dfrac{base}{hypotenuse} = \dfrac{b}{c}$

4. $\sec\theta = \dfrac{hypotenuse}{adjacent\ side} = \dfrac{hypotenuse}{base} = \dfrac{c}{b}$

5. $\tan\theta = \dfrac{opposite\ side}{adjacent\ side} = \dfrac{perpendicular}{base} = \dfrac{a}{b}$

6. $\cot\theta = \dfrac{adjacent\ side}{opposite\ side} = \dfrac{base}{perpendicular} = \dfrac{b}{a}$

7. In several texts, the phrases adjacent side and base, and, opposite side and perpendicular are used interchangeably.

Concept 48: Unit Circle Definition

Consider a Cartesian coordinate system, with O as its origin. We draw a circle of unit radius with its centre at O. A point P, at (x,y) is plotted on the circumference.

We drop a normal from this point to the horizontal diameter or X-axis and the normal intersects the axis at A. The triangle OPA is a right angled triangle. The hypotenuse is equal to the radius, which is of unit length. x denotes the length of base and y denotes the length of the perpendicular. We can write down the trigonometric ratios as below:

1. $\sin\theta = \dfrac{y}{1} = y$

2. $\csc\theta = \dfrac{1}{y}$

3. $\cos\theta = \dfrac{x}{1} = x$

4. $\sec\theta = \dfrac{1}{x}$

5. $\tan\theta = \dfrac{y}{x}$

6. $\cot\theta = \dfrac{x}{y}$

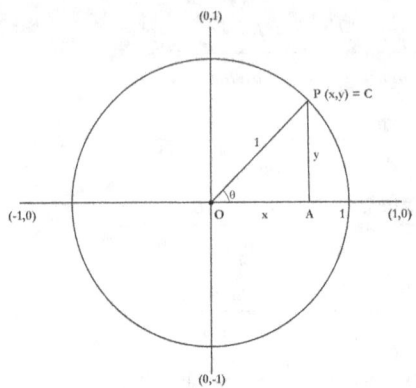

7. We can make a few observations based on the unit circle example we have seen so far.

8. If we had to determine the least positive angle whose sine is equal to a, then we need to drop a perpendicular whose length is a.

9. If we had to determine the least postive angle whose cosine is equal to a, then we ensure that the base is equal to b.

10. The sine function varies from 0 to 1---the perpendicular can be 0, when points A and C are over each other; and the can be one when points A and B overlap. The sine function cannot have a value greater than unity.

11. A similar observation can be made about the cosine function as well.

12. Since tangent is the ratio of sine and cosine functions, the tangent is zero, when sine of an angle is zero. Similarly when cosine is zero, the tangent is not defined. Tangent of an angle is equal to one, when sine and cosine are equal to one another. This happens in the case of a right angles isosceles triangle---where the perpendicular is equal to the base of the triangle.

13. It is easy and informative for us to deduce the basic properties of various ratios based on the unit circle.

14. At this point in time, we are considering angles between 0^0 and 90^0 ---or angles in the first quadrant only.

Concept 49: Heights and Distances

1. Scenario 1
   a. The angle of elevation is the angle formed by a horizontal line, and the line of sight looking up from the horizontal.
   b. The horizontal line under discussion may be real or imagined.

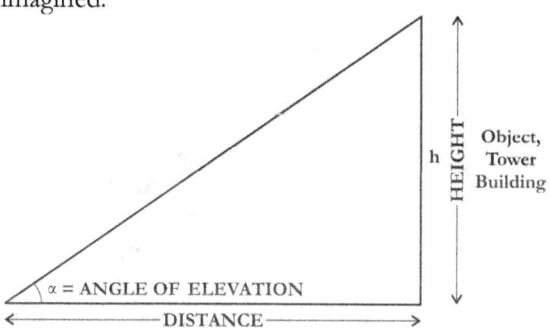

   c. We formulate the problem as a right angle.
   d. We can use the pythogorean theorem to solve for the unknown quantity.
2. Scenario 2
   a. The angle of depression is the angle formed by the horizontal and the line of sight, looking down at the target from a certain elevation.
   b. As before, the horizontal may be real or imagined.

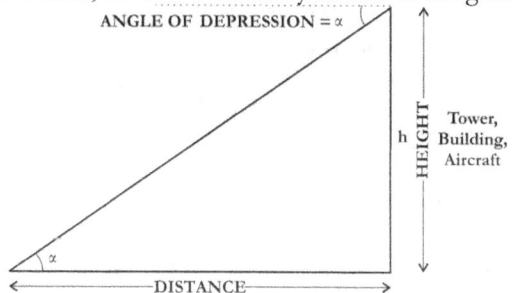

   c. We formulate the representation as a right triangle.
   d. We can use the Pythogorean theorem to solve for the unknown quantity.
3. Scenario 3
   a. When two angles of elevation are given at a specified distance apart, the problem can still be formulated as two right triangles sharing a common elevation and a portion of the horizontal.

97

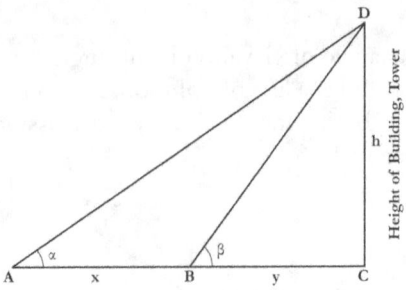

α,β Angle of Elevation from two points A and B

b. These elevation changes can occur in two ways.

1. The observer moves towards the reference elevation resulting in an increase in angle of elevation. This results in a decrease in the horizontal by a delta amount.

2. When the observer moves away from the elevation, the angle of elevation reduces. This results in an increase in the horizontal by a delta amount.

c. We can use the pythogorean theorem to solve for the unknown quantity.

Concept 50: Pythagorean Identitities

1. $\sin^2\theta + \cos^2\theta = 1$

2. $\tan^2\theta + 1 = \sec^2\theta$

3. $1 + \cot^2\theta = \csc^2\theta$

Concept 51: Reciprocal Identities

1. $\csc\theta = \dfrac{1}{\sin\theta}$

2. $\sin\theta = \dfrac{1}{\csc\theta}$

3. $\sec\theta = \dfrac{1}{\cos\theta}$

4. $\cos\theta = \dfrac{1}{\sec\theta}$

5. $\cot\theta = \dfrac{1}{\tan\theta}$

6. $\tan\theta = \dfrac{1}{\cot\theta}$

Concept 52: Trigonometric Ratios in terms of sin, cos and tan

	$\sin\theta$	$\cos\theta$	$\tan\theta$
$\sin\theta$	$\sin\theta$	$\sqrt{1-\cos^2\theta}$	$\dfrac{\tan\theta}{\sqrt{1+\tan^2\theta}}$
$\cos\theta$	$\sqrt{1-\sin^2\theta}$	$\cos\theta$	$\dfrac{1}{1+\tan^2\theta}$
$\tan\theta$	$\dfrac{\sin\theta}{\sqrt{1-\sin^2\theta}}$	$\dfrac{\sqrt{1-\cos^2\theta}}{\cos\theta}$	$\tan\theta$
$\cot\theta$	$\dfrac{\sqrt{1-\sin^2\theta}}{\sin\theta}$	$\dfrac{\cos\theta}{\sqrt{1-\cos^2\theta}}$	$\dfrac{1}{\tan\theta}$
$\sec\theta$	$\dfrac{1}{\sqrt{1-\sin^2\theta}}$	$\dfrac{1}{\cos\theta}$	$\sqrt{1+\tan^2\theta}$
$\csc\theta$	$\dfrac{1}{\sin\theta}$	$\dfrac{1}{\sqrt{1-\cos^2\theta}}$	$\dfrac{\sqrt{1+\tan^2\theta}}{\tan\theta}$

Concept 53: Trigonometric Ratios in terms of cot, sec and cst

	$\cot\theta$	$\sec\theta$	$\csc\theta$
$\cos\theta$	$\dfrac{1}{\sqrt{1+\cot^2\theta}}$	$\dfrac{\sqrt{\sec^2\theta-1}}{\sec\theta}$	$\dfrac{1}{\csc\theta}$
$\tan\theta$	$\dfrac{1}{\cot\theta}$	$\sqrt{\sec^2\theta-1}$	$\dfrac{1}{\sqrt{\csc^2\theta-1}}$

$\cot\theta$	$\cot\theta$	$\dfrac{1}{\sqrt{\sec^2\theta-1}}$	$\sqrt{\csc^2\theta-1}$
$\sec\theta$	$\dfrac{\sqrt{1+\cot^2\theta}}{\cot\theta}$	$\sec\theta$	$\dfrac{\csc\theta}{\sqrt{\csc^2\theta-1}}$
$\csc\theta$	$\sqrt{1+\cot^2\theta}$	$\dfrac{\sqrt{\sec^2\theta-1}}{\sec\theta}$	$\csc\theta$

Concept 54: Tangent and Cotangent Identities

1. $\tan\theta = \dfrac{\sin\theta}{\cos\theta}$

2. $\cot\theta = \dfrac{\cos\theta}{\sin\theta}$

Concept 55: Addition and Subtraction Formulas

1. $\sin(\alpha\pm\beta)=\sin\alpha\cos\beta\pm\cos\alpha\sin\beta$

2. $\cos(\alpha\pm\beta)=\cos\alpha\cos\beta\mp\sin\alpha\sin\beta$

3. $\tan(\alpha\pm\beta)=\dfrac{\tan\alpha\pm\tan\beta}{1\mp\tan\alpha\tan\beta}$

Concept 56: Sum and Product Formulas

1. $\sin\alpha\sin\beta=\dfrac{1}{2}\left(\cos(\alpha-\beta)-\cos(\alpha+\beta)\right)$

2. $\cos\alpha\cos\beta=\dfrac{1}{2}\left(\cos(\alpha-\beta)+\cos(\alpha+\beta)\right)$

3. $\sin\alpha\cos\beta=\dfrac{1}{2}\left(\sin(\alpha+\beta)+\sin(\alpha-\beta)\right)$

4. $\cos\alpha\cos\beta=\dfrac{1}{2}\left(\sin(\alpha+\beta)-\sin(\alpha-\beta)\right)$

5. $\sin\alpha + \sin\beta = 2\sin\left(\dfrac{\alpha+\beta}{2}\right)\cos\left(\dfrac{\alpha-\beta}{2}\right)$

6. $\sin\alpha + \sin\beta = 2\cos\left(\dfrac{\alpha+\beta}{2}\right)\sin\left(\dfrac{\alpha-\beta}{2}\right)$

7. $\cos\alpha + \cos\beta = 2\cos\left(\dfrac{\alpha+\beta}{2}\right)\cos\left(\dfrac{\alpha-\beta}{2}\right)$

8. $\cos\alpha - \cos\beta = -2\sin\left(\dfrac{\alpha+\beta}{2}\right)\sin\left(\dfrac{\alpha-\beta}{2}\right)$

Concept 57: Double Angle Formulas

1. $\sin(2\theta) = 2\sin\theta\cos\theta$

2. $\cos(2\theta) = \cos^2\theta - \sin^2\theta = 2\cos^2 - 1 = 1 - 2\sin^2\theta$

3. $\tan(2\theta) = \dfrac{2\tan\theta}{1 - \tan^2\theta}$

Concept 58: Even and Odd Formulas

1. $\sin(-\theta) = -\sin\theta$
2. $\csc(-\theta) = -\csc\theta$
3. $\cos(-\theta) = \cos\theta$
4. $\sec(-\theta) = \sec\theta$
5. $\tan(-\theta) = -\tan\theta$
6. $\cot(-\theta) = -\cot\theta$

Concept 59: Formulas handling Periodicity

1. $\sin(\theta + 2\pi n) = \sin\theta$
2. $\csc(\theta + 2\pi n) = \csc\theta$
3. $\cos(\theta + 2\pi n) = \cos\theta$
4. $\sec(\theta + 2\pi n) = \sec\theta$
5. $\tan(\theta + \pi n) = \tan\theta$
6. $\cot(\theta + \pi n) = \cot\theta$

Concept 60: Signs of Trigonometric Ratios

The table below captures the change in signs of trigonometric ratios as the magnitude of angles move from one quadrant to another. For visualing this, start with the line in the first quadrant with angle $0°$; and move counterclockwise.

The menmonic for remembering is **ASTC**.

Quadrant 1: **A**ll are positive. None negative

Quadrant 2: **S**ine / cosecant positive; rest negative

Quadrant 3: **T**angent / cotangent positive; rest negative

Quadrant 4: **C**osine / secant positive; rest negative

II Quadrant:	**I Quadrant:**
sin is positive	sin is positive
cos is negative	cos is positive
tan is negative	tan is positive
cot is negative	cot is positive
sec is negative	sec is positive
csc is positive	csc is positive
III Quadrant:	**IV Quadrant:**
sin is negative	sin is negative
cos is negative	cos is positive
tan is positive	tan is negative
cot is positive	cot is negative
sec is negative	sec is positive
csc is negative	csc is negative

Concept 61: Cofunctions

1. $\sin\left(\dfrac{\pi}{2} - \theta\right) = \cos\theta$

2. $\cos\left(\dfrac{\pi}{2} - \theta\right) = \sin\theta$

3. $\csc\left(\dfrac{\pi}{2} - \theta\right) = \sec\theta$

4. $\sec\left(\dfrac{\pi}{2} - \theta\right) = \csc\theta$

5. $\tan\left(\dfrac{\pi}{2} - \theta\right) = \cot\theta$

6. $\cot\left(\dfrac{\pi}{2} - \theta\right) = \tan\theta$

Concept 62: Inverse Trigonometric Functions

1. $y = \sin^{-1} x$ is equivalent to $x = \sin y$
2. $y = \cos^{-1} x$ is equivalent to $x = \cos y$
3. $y = \tan^{-1} x$ is equivalent to $x = \tan y$

Concept 63: Inverse Properties

1. $\cos(\cos^{-1} x) = x$
2. $\cos^{-1}(\cos\theta) = \theta$
3. $\sin(\sin^{-1} x) = x$
4. $\sin^{-1}(\sin\theta) = \theta$
5. $\tan(\tan^{-1} x) = x$
6. $\tan^{-1}(\tan\theta) = \theta$

Concept 64: Laws of Sines, Cosines and Tangents

1. $\dfrac{\sin\alpha}{a} = \dfrac{\sin\beta}{b} = \dfrac{\sin\gamma}{c}$

2. $\dfrac{a-b}{a+b} = \dfrac{\tan\left(\dfrac{\alpha-\beta}{2}\right)}{\tan\left(\dfrac{\alpha+\beta}{2}\right)}$

3. $\dfrac{b-c}{b+c} = \dfrac{\tan\left(\dfrac{\beta-\gamma}{2}\right)}{\tan\left(\dfrac{\beta+\gamma}{2}\right)}$

4. $\dfrac{c-a}{c+a} = \dfrac{\tan\left(\dfrac{\gamma-\alpha}{2}\right)}{\tan\left(\dfrac{\gamma+\alpha}{2}\right)}$

5. $a^2 = b^2 + c^2 - 2bc\cos\alpha$

6. $b^2 = a^2 + c^2 - 2ac\cos\beta$

7. $c^2 = a^2 + b^2 - 2ab\cos\gamma$

Concept 65: Analytical Trigonometry

1. **Trigonometric expressions**, contain trigonometric terms and relationships.
2. **Trigonometric equations**, contain a trigonometric expression followed by an "equal to" sign; these can be solved just like one would in the case of algebraic expressions.
3. All algebraic techniques like brackets, factorization etc can be applied. In some cases, we substitute an algebraic variable for a trigonometric term. In doing so, the problem gets converted to a pure algebraic expression.
4. Students are advised to understand the trigonometric identities and commit them to memory.

Concept 66: Common Trigonometric Ratios

The following table captures common trigonometric ratios that we often come across while solving problems. With practice, you will be able to recall these values without much effort. While you review the table, you will see several patterns. Let us spend a moment to notice and record a few patterns.

Deg	0	30	45	60	90	120	135	150	180

Rad	0	$\dfrac{\pi}{6}$	$\dfrac{\pi}{4}$	$\dfrac{\pi}{3}$	$\dfrac{\pi}{2}$	$\dfrac{2\pi}{3}$	$\dfrac{3\pi}{4}$	$\dfrac{5\pi}{6}$	2π
sin	0	$\dfrac{1}{2}$	$\dfrac{1}{\sqrt{2}}$	$\dfrac{\sqrt{3}}{2}$	1	$\dfrac{\sqrt{3}}{2}$	$\dfrac{1}{\sqrt{2}}$	$\dfrac{1}{2}$	0
cos	1	$\dfrac{\sqrt{3}}{2}$	$\dfrac{1}{\sqrt{2}}$	$\dfrac{1}{2}$	0	$\dfrac{-1}{2}$	$\dfrac{-1}{\sqrt{2}}$	$\dfrac{-\sqrt{3}}{2}$	-1
tan	0	$\dfrac{1}{\sqrt{3}}$	1	$\sqrt{3}$	∞	$\sqrt{3}$	-1	$\dfrac{-1}{\sqrt{3}}$	0
cot	∞	$\sqrt{3}$	1	$\dfrac{1}{\sqrt{3}}$	0	$\dfrac{-1}{\sqrt{3}}$	-1	$-\sqrt{3}$	∞
csc	∞	2	$\dfrac{1}{\sqrt{2}}$	$\dfrac{2}{\sqrt{3}}$	1	$\dfrac{2}{\sqrt{3}}$	$\sqrt{2}$	2	∞
sec	1	$\dfrac{2}{\sqrt{3}}$	$\sqrt{2}$	2	∞	-2	$-\sqrt{2}$	$\dfrac{2}{\sqrt{3}}$	-1

1. The sin and cos rows are similar but in reverse order.
2. The tan and cot are reciprocals. Corresponding entries are reciprocals as well.
3. The sec and csc are simply reciprocals of corresponding values of cos and sin.

Closing Thoughts

There are five fundamental principles, or say **good habits** that we would like to emphasize before we commence our discussion on Mathematics.

1. Neatness is conducive to accuracy. Refrain from the temptation to write down something quickly and then scratch the same to make the necessary corrections.

2. One of the weaknesses we find in students while solving word problems is the usage of = sign. This sign has a specific meaning in the world of mathematics. It cannot be used as a way to begin every new line or step in the problem solving process. Use appropriate mathematical signs and symbols. Never use them to mean something vague. = sign is not a space-filler.

3. Spend a second or two to explain how you arrived at a certain step. Several books and references use a statement, such as "it follows from the above statement". We have oftentimes wondered how the expression or equation below follows from the one above. A good explanation is an excellent demonstration of your understanding of the underlying principles.

4. When you are faced with several conclusions during problem solving process, it is a good idea to number the statements or equations. In subsequent steps, you can refer to these conclusions by using the label or the assigned equation number.

5. The easiest of problems attracts the silliest of mistakes. If the problem is easy, motivate yourself to get it right. Do not let over-confidence or carelessness take control of the situation.

www.ingramcontent.com/pod-product-compliance
Lightning Source LLC
Chambersburg PA
CBHW072036190526
45165CB00017B/947